# String Fields, Higher Spins and Number Theory

**Other Related Titles from World Scientific**

---

*Higher Spin Gauge Theories*
edited by Lars Brink, Marc Henneaux and Mikhail A Vasiliev
ISBN: 978-981-3144-09-5

*Theory of Groups and Symmetries: Finite Groups, Lie Groups, and Lie Algebras*
by Alexey P Isaev and Valery A Rubakov
ISBN: 978-981-3236-85-1

*From Fields to Strings: Circumnavigating Theoretical Physics: Ian Kogan Memorial Collection (In 3 Volumes)*
edited by Misha Shifman, Arkady Vainshtein and John Wheater
ISBN: 978-981-238-955-8 (Set)
ISBN: 978-981-256-000-1 (Vol. 1)
ISBN: 978-981-256-001-8 (Vol. 2)
ISBN: 978-981-256-114-5 (Vol. 3)

# String Fields, Higher Spins and Number Theory

**Dimitri Polyakov**

*Center for Theoretical Physics, Sichuan University, China*
*Institut des Hautes Etudes Scientifiques (IHES), France*

NEW JERSEY • LONDON • SINGAPORE • BEIJING • SHANGHAI • HONG KONG • TAIPEI • CHENNAI • TOKYO

*Published by*

World Scientific Publishing Co. Pte. Ltd.
5 Toh Tuck Link, Singapore 596224
*USA office:* 27 Warren Street, Suite 401-402, Hackensack, NJ 07601
*UK office:* 57 Shelton Street, Covent Garden, London WC2H 9HE

**Library of Congress Cataloging-in-Publication Data**
Names: Polyakov, Dmitrii (Dmitrii Borisovich), author.
Title: String fields, higher spins, and number theory / by Dimitri Polyakov
(Sichuan University, China).
Description: New Jersey : World Scientific, 2018. | Includes bibliographical references.
Identifiers: LCCN 2018039784| ISBN 9789813233393 (hardcover : alk. paper) |
ISBN 9813233397 (hardcover : alk. paper)
Subjects: LCSH: String models. | Nuclear spin. | Particles (Nuclear physics) | Number theory.
Classification: LCC QC794.6.S85 P668 2018 | DDC 539.7/258--dc23
LC record available at https://lccn.loc.gov/2018039784

**British Library Cataloguing-in-Publication Data**
A catalogue record for this book is available from the British Library.



For any available supplementary material, please visit
https://www.worldscientific.com/worldscibooks/10.1142/10800#t=suppl

Desk Editor: Ng Kah Fee

Typeset by Stallion Press
Email: enquiries@stallionpress.com

Printed in Singapore

*Dedicated to the memory of Ian Kogan, my wonderful friend and collaborator*

# Contents

# Preface

In this book, I describe (in a very incomplete and intuitive manner) some intriguing interplays existing between Higher-Spin Gauge Theories String Theory and String Field Theory, with the Number Theory (in particular, the Theory of Partitions) often appearing as a natural interface. All of these subjects are immense, deep and fascinating, all of them contain enormous number of hot topics that have been at the cutting edge of research in mathematical physics and high-energy physics over recent years. This book definitely does not aim to overview any of these subjects in a self-contained manner; there exists immense amount of excellent literature on all of them. My purpose in this book was far more modest: it is a summary of my attempts to explore interplays and relations between string theory in various space-time backgrounds and higher-spin gauge theories by using a very special string-theoretic approach, emerging from the concept of "larger string theory", explained in the book. This concept is somewhat similar to the concept of background independence in string field theory and implies that string theory, originally defined around flat space-time background, actually posseses hidden AdS space-time isometries, as well as higher-spin envelopings of these isometries. The operators realizing these symmetries can be classified in terms of ghost cohomologies, which appearance in string theory is related to peculiar geometrical properties of supermoduli spaces in Ramond-Neveu-Schwarz superstring theory. Using these symmetry generators, one can construct vertex operators describing emissions of higher-spin modes (including the massless ones) by open and closed strings and compute the correlation functions describing interactions of higher-spin fields in AdS geometry. The appearance of the AdS geometry is important in our scenario, since

1) this geometry is crucial for the holography principle (including higher-spin holography), relating higher-spin interactions to conformal blocks in the dual theories;

2) it is the geometry where the consistent formulation of higher-spin theory is possible by circumventing the restrictions of Coleman-Mandula theorem.

In turn, in string field theory (SFT), relations between different space-time backgrounds can be realized through analytic solutions of SFT equations of motion, interpolating between BRST cohomologies describing string dynamics in various space-time geometries. Although the complete SFT solution interpolating between flat and AdS geometries have not yet been found, some efforts have been made in this direction and are described in this book. There are many hints at SFT and higher-spin theories being related since the underlying equations of these theories (Vasiliev's equations and SFT equations of motion) are structurally quite similar. Making this similarity explicit is, however, a hard problem due to the complexity of the star products appearing in both of them. In this book we explore this question and describe classes of analytic solutions in SFT, relevant to higher-spin dynamics in $AdS$. The study of the structures of higher-spin interactions and of the analytic solutions in SFT points out some common properties: the appearance of rather complex summations over partitions of integers and of the Bell polynomial operators that, on the one hand, can be regarded as the elementary building blocks realizing higher-spin symmetries in terms of operator algebras and, at the same time, relevant to the important and long-standing number-theoretical problem of counting the numbers of partitions of integers. This book culminates in describing the solution of this problem by using the operator product techniques involving the irregular vertex operators that can be thought as "generating wavefunctions" for higher-spin modes in string theory. The book is organized as follows. In Chapter 1 we briefly review the connections between the strings and higher-spin dynamics in the traditional approach, using the tensionless limit of string theory. This approach is limited to strings propagating in flat space-time background. There are many complications related to this approach, notably, this is the limit opposite to the low-energy limit of string theory, making it hard to relate string-theoretic correlators, computed in the tensionless limit, to gauge-invariant interaction terms in the field-theoretic effective action. The approach described in this book, is quite different and is developed directly for the $\alpha' \to 0$ limit, opposite to the tensionless case. In Chapter 2 we explain the geometrical aspects, paving the ground for the "larger string theory", based on ghost picture inequivalence of certain sectors of physical states in RNS superstring theory, by analyzing geometrical singularities in supermoduli space of string amplitudes. This inequivalence can be described and classified in terms of ghost cohomologies $H_n \sim H_{-n-2}$ that turn out to realize hidden symmetries in the target space. These symmetries are analyzed in the Chapter 3. It is shown that the generators from $H_1 \sim H_{-3}$ ($n = 1$) realize AdS space-time isometry algebra, while the generators of the higher cohomologies realize the infinite-dimensional higher-spin enveloping of this algebra. The fusion rules for the cohomologies are identical to the structure of the $AdS$ higher-spin algebras with the spin value of the currents and the order $n$ related as $s = n + 2$. Based on these symmetry generators, in Chapter 4 we construct vertex operators describing emissions of massless higher-spin particle by a string, which are also classified by $H_1 \sim H_{-3}$. In Chapter 4, using this construction, we discuss the emergence of $AdS$ geometry in space-time relating it to the nonzero ghost cohomology sectors of superstring theory and compute concrete examples of cubic, quartic and quintic higher-spin interactions. The nonlocality of the interactions is also related to the nontrivial ghost structure of the operators. This naturally leads us to the idea of background independence, which is most naturally formulated in terms of analytic SFT solutions, considered in Chapter 5. The class of analytic string field theory solutions, considered in this chapter, parametrizes higher-spin symmetry algebra in $AdS_3$ and may be considered a candidate to describe a collective non-perturbative higher-spin vacuum, in the same spirit that Schnabl's solutions describe the nonperturbative tachyonic vacuum in bosonic string theory. The results described in this chapter are incomplete, as they only define the BRST cohomologies describing this hypothetic vacuum state. Explicit analysis of these cohomology structures, as well as their relation to higher-spin holography, is left for the future. The analytic SFT solution describing the flat-to-AdS deformations is still missing, although, based on general holographic arguments, its structure is likely to involve the irregular vertex operators, studied in Chapters 6 and 7. The book culminates in proposing the explicit solution to the problem of counting numbers of restricted partitions of fixed lengths in terms of finite series, which is an important long-standing problem in number theory. The solution is obtained by computing two-point CFT correlators of specific irregular vertex operators, counting the partition numbers, and is given by finite series in regularized higher-order Schwarzians of a certain conformal transformation. All these results demonstrate intriguing interplays existing between strings, holography and number theory, considered in this book. These results are obviously just a tiny tip of an iceberg.

In conclusion, I can't find enough words to express my deep gratitude to my colleagues, my parents Alexander and Marina, my close friends (with some of them unfortunately not with us anymore), whose vision, help, advice and friendship inspired and helped me all the way through. This book is wholeheartedly dedicated to the memory of Ian Kogan, my wonderful friend and collaborator. Many of the ideas described in this book started to crystallize in our inspiring and illuminating conversations and projects, interrupted by his untimely tragic death in 2003. I would like to thank all the people in the higher spin and the string field theory communities for great discussions and the organizers of conferences and workshops on these subjects which were always a great inspiration with new illuminating insights; in particular, it is my great pleasure to thank Massimo Bianchi, Loriano Bonora, Nicholas Boulanger, Lars Brink, Slava Didenko, Dario Francia, Marc Henneaux, Nobuyuki Ishibashi, Euihun Joung, Yoshi Kitazawa, Igor Klebanov, Bum-Hoon Lee, Tetsuya Masuda, Hermann Nicolai, Bo Ning, Rakib Rahman, Soo-Jong Rey, Chaiho Rim, Augusto Sagnotti, Zhenia Skvortsov, Zheng Sun, Per Sundell, Misha Vasiliev, Haitang Yang and many, many others - the list could be continued indefinitely. I also wish to express my deep gratitude to the World Schientific Publishers, in particular to the Editor Mr. Ka-Fee Ng for the patience, guidance and encouragement. It is also very personal to express my gratitude to Daria Oleinikova and Sasha Voloshin, both are my very special friends, whose encouragement and support is always inspiring and heart-warming.

*Dimitri Polyakov*

# Chapter 1

# Introduction: Massive Higher-Spin Modes in String Dynamics and Tensionless Limit

The holographic principle (with the AdS/CFT conjecture being one of its manifestations) is a deep and fascinating physical law, holding keys to some profound mysteries of how our Universe functions. This principle connects seemingly unrelated physical theories, living in different space-time dimensions with very different geometries. These theories altogether vary from string theory and gauge theories to models in condensed matter physics, as well as theories of turbulence and chaos. String theory, however, is "more equal than others" in this list and appears to be pivotal in the whole picture. For many years string theory was hoped to be one of the leading candidates for grand unification of fundamental interactions. Although today we know that those hopes were, in a sense, too optimistic and string theory isn't likely to ever become a "theory of everything", it still remains our best hope to put gravity in the context of quantum mechanics and quantum field theory, as well as to understand true symmetries of space-time. At the same time, string theory developed what can be seen as unique and appropriate language to approach holography, AdS/CFT correspondence and dualities between gravity and gauge/conformal field theories in various dimensions. For example, it is common to think of AdS/CFT correspondence as of relation between gravity with negative cosmological constant (which vacuum solution is anti de Sitter geometry) and conformal field theory living on the asymptotically flat boundary of such a solution. However, this is only the low-energy approximation of the AdS/CFT duality. In the complete sense, the AdS/CFT correspondence, if it is true, must imply the exact isomorphism between vertex operators in string theory (in anti de Sitter space-time) and gauge (conformally) invariant observables (operators) on the AdS boundary. Namely, the worldsheet correlation functions of vertex operators, computed in $D$-dimensional AdS string theory, must reproduce the correlators of the corresponding dual operators in $D-1$ dimensional conformal field theory. For example, the $N$-point correlation function of dilaton vertex operators in $AdS_5$ string theory (with the dilatons propagating parallel the AdS boundary) must reproduce the $N$-point correlator of glueball operators $< Tr(F^2)...Tr(F^2) >$ in $D=4$ Yang-Mills theory. This picture immediately leads to several observations. First of all, it is clear that the $AdS$ duals of most of the operators on the conformal field theory side (such as composite operators) are inevitably mixed-symmetry higher-spin fields, propagating in the AdS bulk. Thus the higher-spin dynamics in AdS space is the first crucial ingredient, necessary to describe holography. The second ingredient is string theory in $AdS$ space. Here the major difficulty is the background dependence of the first-quantized string theory. Equations of motion of strings in curved backgrounds are essentially nonlinear, with no canonical or BRST quantizations no worldsheet CFT descriptions existing for such theories, apart from string perturbation theory. In particular, little is known about string dynamics in anti de Sitter spaces beyond semi-classical limit, in which very limited examples of correlation functions can be computed. This implies that the string field theory (second-quantized string theory) may be much more natural a language to describe the holography from string-theoretic point of view, in particular, making it possible explore strings in curved backgrounds, such as AdS. Here, the key property of string field theory is the background independence. Thus string field theory and higher-spin gauge theories emerge as two crucial building blocks of the holography principle, although the significance of these two fascinating subjects extends far beyond this principle.What is even more remarkable, the fundamental equations describing string field theory and higher-spin gauge theory turn out to be strikingly similar. Is it more much than just a formal similarity? Are the connections existing between these two subjects actually related to fundamentals of holography and gauge-string correspondence and far beyond? How does the number theory enter the game? Our purpose will be to investigate these questions in what follows.

## 1.1 String Theory and Higher-Spin Dynamics

As noted above, higher-spin gauge fields are crucial in the holographic context, as they appear as $AdS$ gravity duals of huge variety of composite operators on the CFT side. There are also many other reasons why higher-spin gauge theories are of great interest to us. Although it may happen that one may not directly observe the higher-spin particles in accelerators, higher-spin gauge symmetries, extending the space-time symmetries, are crucial to understand the true symmetries of early Universe and unification models. In this context, one may think of a conventional string theory (with the Einstein gravity being its low-energy limit) as of a spontaneously broken phase of a larger theory, involving the high-spin dynamics. The generators of the higher-spin algebra (outside its maximal finite subalgebra) should then be related to the "Goldstone modes" of such a breaking. One of the deep problems in conventional string theory is that string perturbation theory by itself does not tel us anything about the vacuum configuration of the target space. The key to understanding the structure of the string vacuum is then possibly hidden in the unbroken theory that is, consistently interacting higher-spin gauge theories. These theories are known to be profoundly hard to formulate. This is the fascinating and complicated problem that has attracted a profound interest over many years since the 30s. Higher-spin field theory is an example of how a deep and profound discovery can emerge as a result of a "wrong" motivation. It is well-known how Yukawa has predicted an existence of a new particle (a meson) and calculated its mass with astounding precision, which ultimately lead to complete overhaul and breakthrough in particle physics. However, this discovery has been made for a "wrong" reason, as Yukawa was attempting to describe a particle mediating strong interaction. In a sense, the history of higher-spin gauge theories initiated in a similar spirit: Dirac was one of the first to study the higher-spin objects in order to describe neutrino physics. And, even though we know today that neutrino is not a higher-spin particle, and higher-spin theories have been thought to be inconsistent and irrelevant to physics, this unorthodox study has ultimately led to what turned out to be a highly fascinating subject. Despite the fact that Pauli and Fierz succeeded in finding suitable on-shell conditions for symmetric higher-spin fields, the action principle for such fields has only been formulated by Fronsdal in 1978. In $D$-dimensional Minkowski space-time, the gauge-invariant action for massless symmetric spin $s$ field $\varphi_{m_1\ldots m_s}$ is

$$\begin{aligned}
S = \frac{1}{2}\int d^D x[&\partial_n\varphi_{m_1\ldots m_s}\partial^n\varphi^{m_1\ldots m_s} \\
&-\frac{s(s-1)}{2}\partial_n\varphi^p_{pm_3\ldots m_s}\partial^n\varphi_r^{rm_3\ldots m_s} \\
&+s(s-1)\partial_n\varphi^r_{rm_3\ldots m_s}\partial_p\varphi_r^{npm_3\ldots m_s} - s\partial_n\varphi^n_{m_2\ldots m_s}\partial_p\varphi^{pm_2\ldots m_s} \\
&-\frac{s(s-1)(s-2)}{2}\partial_n\varphi^{np}_{pm_4\ldots m_s}\partial_r\varphi_q^{rqm_4\ldots m_s}]
\end{aligned} \tag{1.1}$$

where $\varphi$ is double-traceless:

$$\eta^{m_1m_2}\eta^{m_3m_4}\varphi_{m_1...m_s}=0. \tag{1.2}$$

This action is invariant under gauge transformation:

$$\delta\varphi_{m_1...m_s}=\partial_{(m_1}\epsilon_{m_2...m_s)} \tag{1.3}$$

where (...) denotes symmetrization, and the completely symmetric spin $s-1$ gauge parameter is traceless:

$$\eta^{m_1m_2}\Lambda_{m_1m_2...m_{s-1}}=0. \tag{1.4}$$

For spin $s=2$ the action (1.1) is the linearized Einstein-Hilbert action and the gauge transformations (1.3) are the linearized diffeomorphisms. In the AdS case, the gauge-invariant action has the form:

$$\begin{aligned}S=\frac{1}{2}\int d^Dx\sqrt{g_{AdS}}\{&g^{mn}_{AdS}\nabla_m h^{m_1...m_s}\nabla_n h_{m_1...m_s}\\&-s\nabla_m h^{mm_2...m_s}\nabla^n h_{nm_2...m_s}\\&+s(s-1)\nabla^n h^{mm_3...m_s}_m\nabla^p h_{pnm_3...m_s}\\&-\frac{s(s-1)}{2}g^{mn}_{AdS}\nabla_m h^{pm_3...m_s}_p\nabla_n h^q_{qm_3...m_s}\\&-\frac{s(s-1)(s-2)}{4}\nabla_p h^{qpm_4...m_s}_q\nabla^r h^s_{rsm_4...m_s}\\&-\rho(s^2-s-2)h_{m_1...m_s}h^{m_1...m_s}+\frac{\rho}{2}h^{pm_3...m_s}_p h^q_{qm_3...m_s}\}\end{aligned} \tag{1.5}$$

where the $AdS_D$ metric is

$$\begin{aligned}&ds^2=\frac{R^2}{z^2}(dz^2+dx_t dx^t)\\&0\le t\le D-2\end{aligned} \tag{1.6}$$

and $\rho=-\frac{1}{R^2}$ is the cosmological constant. Note the appearance of the mass-like term in the action (1.5). It is particularly related to the structure of covariant derivative in the AdS space: $\nabla\sim\partial+R^{-1}$ and reflect some very special properties of vertex operators for higher spins in string theory that will be discussed in the next chapters. Attempting to introduce the gauge-invariant higher-spin interactions makes things increasingly complex. Despite strong efforts by some leading experts in recent years [1–3, 11–17, 19–22, 26–29, 37–48] there are still key issues about these theories that remain unresolved (even for the non-interacting particles; much more so in the interacting case).

There are several reasons why the higher spin theories are so complicated. First of all, in order to be physically meaningful, these theories need to possess sufficiently strong gauge symmetries, powerful enough to ensure the absence of unphysical (negative norm) state s. For example, in the Fronsdal's description [4] the theories describing symmetric tensor fields of spin $s$ are invariant under gauge transformations with the spin $s-1$ traceless parameter. Theories with the vast gauge symmetries like this are not trivial to construct even in the non-interacting case,when one needs to introduce a number of auxiliary fields and objects like non-local compensators [2,3,23–25] Moreover, as the gauge symmetries in higher spin theories are necessary to eliminate the unphysical degrees of freedom, they must be preserved in the interacting case as well, i.e. one faces a problem (even more difficult) of introducing the interactions in a gauge-invariant way. In the flat space things are further complicated because of the no-go theorems (such as Coleman-Mandula theorem [5, 6]) imposing strong restrictions on conserved charges in interacting theories with a mass gap, limiting them to the scalars and those related to the standard Poincare generators. Thus Coleman-Mandula theorem in $d = 4$ makes it hard to construct consistent interacting theories of higher spin, at least as long as the locality is preserved, despite several examples of higher spin interaction vertices constructed over the recent years [14, 16, 18]. In certain cases, such as in AdS backgrounds, the Coleman-Mandula theorem can be circumvented (since there is no well-defined S-matrix in the AdS geometry) and gauge-invariant interactions can be introduced consistently - as it has been done in the Fradkin-Vasiliev construction [8–13]. In non-AdS geometries, however (such as in the flat case), the no-go theorems do lead to complications, implying, in particular, that the interacting gauge-invariant theories of higher spins have to be essentially non-local and/or non-unitary. Despite that, some examples of gauge-invariant cubic higher-spin vertices have been known for a long time and, at present, the cubic higher-spin interactions for totally symmetric fields in flat and AdS spaces are mostly classified and higher-spin theories are known to be consistent at the cubic order. From the holographic point of view, these interactions determine the structure constants in dual conformal field theories, living on the AdS boundary. At higher order, things become ever more complicated. The vertices at quartic and higher orders are highly nonlocal, with complicated structures of the nonlocalities. At present, no classification for such vertices is known. The case of the fields with the mixed symmetry is understood even less, with even the non-interacting case in AdS still beyond our grasp. In the following sections, we will review and explained how the string theory approach can help to address all these problems.

## 1.2 Massive Higher Spin Modes in Bosonic String Theory: A Brief Review

To understand how the string theory techniques work for higher-spin gauge theories, let us start with a warm-up example of standard bosonic string theory. Higher spin modes appear naturally as massive excitations in open and closed string theories. For simplicity, consider the open strings (critical or noncritical). On the leading Regge trajectory, the excitations of mass $m$ have the the spin $s$ such that $m \sim \sqrt{2(s-1)}$ and are described by the vertex operators:

$$V = H_{m_1 \dots m_s}(p) \oint dz \partial X^{m_1} \dots \partial X^{m_s} e^{ipX}(z) \tag{1.7}$$

and

$$V = H_{m_1 \dots m_s}(p) c \partial X^{m_1} \dots \partial X^{m_s} e^{ipX}(z) \tag{1.8}$$

in integrated and unintegrated forms respectively, with the integral takem over the worldsheet boundary. The higher-spin field $H_{m_1 \dots m_s}(p)$ is totally symmetric here by construction, as it is obvious from the structure of the vertex operators (1.9). With the BRST charge

$$\begin{aligned} Q &= \oint \frac{dz}{2i\pi}(cT - bc\partial c)(z) \\ T(z) &= \frac{1}{2}\partial X_m \partial X^m + 2\partial c b - \partial b c \end{aligned} \tag{1.9}$$

with $T(z)$ being the full matter+ghost stress-energy tensor, the BRST-invariance condition on $V$: $[Q, V] = 0$ entails the on-shell conditions on $H$:

$$\begin{aligned} &(p^2 - 2(s-1))H_{m_1 \dots m_s}(p) = (p^2 - m^2)H_{m_1 \dots m_s}(p) = 0 \\ &p^{m_1} H_{m_1 \dots m_s}(p) = 0 \\ &\eta^{m_1 m_2} H_{m_1 m_2 \dots m_s}(p) = 0 \end{aligned} \tag{1.10}$$

which are the Pauli-Fiertz on-shell conditions. The gauge transformations for higher-spin fields in space-time are those that shift the vertex operators by BRST-exact terms that do not contribute to correlation functions computed on the worldsheet. The higher-spin interactions, read off those correlators, are then gauge-invariant by construction. This is what makes the string-theoretic approach so powerful: in the string theory language the problem of gauge-invariance of the higher-spin interactions is translated into the one of BRST-properties (invariance and non-triviality) of the vertex operators, which is conceptually far easier. The vertex operator (1.7), however, does not by itself possess any manifest gauge symmetry, as except for the $s = 1$ case, the gauge transformations of the type (1.3) shift it by terms which are clearly not BRST-exact, as it is straightforward to check (e.g. by computing any correlation function with the shift term insertion). This is quite predictable since the operator (1.7) is massive for $s \neq 1$ but again raises the issue of the negative norm modes, making the theory unphysical. To resolve this problem, introduce the traceless symmetric spin $s-1$ Stueckelberg field $B_{m_1...m_{s-1}}$ and deform the vertex operators (1.7), (1.8) according to

$$\begin{aligned}
&V_{s;integrated}(p) = \oint \frac{dz}{2i\pi} e^{ipX}(\partial X^{m_1}...\partial X^{m_s} H_{m_1...m_s}(p) \\
&+s(s-1)\partial X^{m_1}...\partial X^{m_{s-2}}\partial^2 X^{m_{s-1}} B_{m_1...m_{s-1}}(p)) \\
&V_{s;unintegrated}(p) = e^{ipX}(\partial X^{m_1}...\partial X^{m_s} H_{m_1...m_s}(p) \\
&+s(s-1)\partial X^{m_1}...\partial X^{m_{s-2}}\partial^2 X^{m_{s-1}} B_{m_1...m_{s-1}}(p))
\end{aligned} \tag{1.11}$$

The BRST-invariance of $V$ requires

$$ip^{m_s} H_{m_1...m_s} = 2(s-1) B_{m_1...m_{s-1}} \tag{1.12}$$

The Stueckelberg symmetry transformations:

$$\begin{aligned}
&\delta H_{m_1..m_s} = ip_{(m_s} \Lambda_{m_1...m_{s-1})} \\
&\delta B_{m_1...m_{s-1}} = \Lambda_{m_1...m_{s-1}}
\end{aligned} \tag{1.13}$$

leave the integrated vertex operator invariant (shifting the integrand by total derivative) and shift the unintegrated vertex by BRST-exact term

$$\delta V_{unintegrated} = [Q, \Lambda_{m_1...m_{s-1}} e^{ipX} \partial X^{m_1}...\partial X^{m_{s-1}}] \tag{1.14}$$

thus "restoring" the broken gauge symmetry. It is then straightforward to calculate the gauge-invariant 3-vertex for arbitrary massive symmetric higher-spin fields $s_1$, $S_2$, $s_3$ for the leading trajectory. In this case, take all the vertex operators $V_{s_1,s_2,s_3}$ unintegrated at points 0, 1 and $\infty$. Using the BRST symmetry, it is convenient to transform the Stueckelberg term according to:

$$\begin{aligned}
&c\partial X^{m_1}...\partial X^{m_{s-2}}\partial^2 X^{m_{s-1}} e^{ipX} B_{m_1...m_{s-1}}(p) \\
&= -\frac{i}{s(s-1)} c\partial X^{m_1}...\partial X^{m_s} e^{ipX} F_{m_1...m_s}(p) \\
&+[Q, \partial X^{m_1}...\partial X^{m_{s-1}} e^{ipX} B_{m_1...m_{s-1}}]
\end{aligned} \tag{1.15}$$

where

$$F_{m_1...m_s} = ip_{(m_s}B_{m_1...m_{s-1})} \tag{1.16}$$

Now it is convenient to introduce the partitions:

$$\begin{aligned} s_1 &= t_{12} + t_{13} + u_{12} + u_{13} \\ s_2 &= t_{21} + t_{23} + u_{21} + u_{23} \\ s_3 &= t_{31} + t_{32} + u_{31} + u_{32} \\ t_{ij} &= t_{ji} \end{aligned} \tag{1.17}$$

where $t_{ij} = t_{ji}$ count the couplings between derivatives of $X$ of vertex operators for $s_i$ and $s_j$ $(i, j = 1, 2, 3)$ and $u_{ij}$ count the couplings between $X$-derivatives of $s_i$-operator and the exponent of $s_j$-operator (with each of those producing the factor proportional to the momentum of $s_j$-operator. Then the straightforward calculation of the three-point function gives the higher-spin 3-vertex:

$$\begin{aligned} &< V_{s_1}(k;0)V_{s_2}(p;1)V_{s_3}(q;\infty) > \\ &= -\sum_{s_1,s_2,s_3|\{t_{ij}\};\{u_{ij}\}} \{\frac{s_1!s_2!s_3!}{\prod_{1\leq i<j\leq 3} t_{ij}!s_{ij}!s_{ji}!} \\ &\times (ik)^{c_1}....(ik)^{c_{u_{21}}}(ik)^{e_1}...(ik)^{e_{u_{31}}} \\ &\times (-ip)^{a_1}...(-ip)^{a_{u_{12}}}(ip)^{f_1}...(ip)^{f_{u_{32}}} \\ &\times (-iq)^{b_1}...(-iq)^{b_{13}}(-iq)^{d_1}...(-iq)^{d_{u_{23}}} \\ &\times G_{m_1...m_{t_{12}}r_1...r_{t_{13}}a_1...a_{u_{12}}b_1...b_{u_{13}}}(k) \\ &\times G^{m_1...m_{t_{12}}}_{n_1...n_{t_{23}}c_1...c_{u_{21}}d_1...d_{u_{23}}}(p) \\ &\times G^{n_1...n_{t_{23}}r_1....r_{t_{13}}}_{e_1...e_{u_{31}}f_1...f_{u_{32}}}(q)\} \end{aligned} \tag{1.18}$$

where

$$G_{m_1...m_s}(p) = H_{m_1...m_s}(p) - ip_{(m_s}B_{m_1...m_{s-1})} \tag{1.19}$$

and the summations $\Sigma^{(0)}$ and $\Sigma^{(1)}$ taken over all the non-ordered partitions of $s_1$, $s_2$ and $s_3$ with that satisfy the constraint:

$$2t_{13} + 2t_{23} + u_{13} + u_{31} + u_{23} + u_{32} = s_3 \tag{1.20}$$

The last constraint stems from the conformal and BRST invariance of the amplitude (1.18) that only allow the terms behaving asymptotically as $\sim {z_3}^0$ as the location of the third operator is taken to infinity: $\sim z_3 \to \infty$.

In the similar way, correlation functions of open string theory can be used to extract interaction vertices for the states on subleading trajectories, including those with the mixed symmetries.

The structure constants, defined by the tree-point functions, define the leading order contributions to the worldsheet $\beta$-functions and, accordingly, the cubic higher-spin vertices in the low-energy effective action. One can furthermore study the higher-point correlator extracting the $\alpha'$ corrections to the $\beta$-functions, defining the higher order spin interactions (e.g. at quartic, quintic or even higher orders).

Despite that, the bosonic open string theory largely remains not so realistic a toy model of the higher-spin interactions, for several reasons: first, all the higher-spin excitations are massive with masses of the order of $\sim (\alpha')^{-\frac{1}{2}}$, making them physically irrelevant, unless one considers the tensionless limit, $\alpha' \to \infty$. At the same time, the tensionless limit is opposite to the one of the string perturbation theory. For this reason, it remains largely obscure how to connect the correlators, computed in the tensionless limit, to the interaction terms in the low-energy effective action. Second, as was described above, higher-spin interactions (at least, beyond the cubic order) must be essentially nonlocal. However, standard string-theoretic amplitudes (such as Veneziano amplitude) bear no traces of that, all leading to perfectly local terms in the low-energy action. In this sense, standard string theory does not distinguish between lower and higher spins. This is related to the fact that, standard bosonic string theory can be considered as a broken symmetry phase of the full higher-spin theory.In bosonic string theory, the space-time isometries are simply the standard Poincare transformations, generated by operators:

$$T^m = \oint \frac{dz}{2i\pi} \partial X^m$$
$$T^{mn} = \oint \frac{dz}{2i\pi} \partial X^{[m} X^{n]} \quad (1.21)$$

In string theory, the currents for the Poincare generators are the primary fields of dimension 1, which worldsheet integrals are the elements of the BRST cohomology. Given the space-time isometry algebra $G$, the higher-spin algebra can be realized as the enveloping of $G$, e.g. in terms of Weyl-ordered polynomials in the generators of $G$, involving the normally ordered products of the currents.

In bosonic string theory all such products have conformal dimensions greater than one and therefore are either BRST-exact or not BRST-closed, indicating that the absense of any unbroken higher-spin symmetries on-shell bosonic string theory. Can one think, however, of an existence of a certain "larger" string theory, holding the keys to the unbroken phase of higher-spin symmetries as well? The "standard" string theory does not contain any massless higher-spin excitations. In the theory with the full unbroken symmetry, can one observe the whole tower of massless higher-spin excitations? What about the AdS background, crucial for the higher-spin holography? In the following chapters we will address these questions. The first hint comes from the geometric structure of superstring amplitudes in Neveu-Ramond-Schwarz (NRS) superstring theory.

# Chapter 2

# Geometry of Superconformal Moduli in Superstring Theory

## 2.1 Bosonic and Fermionic Ghost Pictures in String Amplitudes

It turns out that, in order to explore the emergence of *AdS* isometries and *AdS* higher-spin symmetries, it is far more natural easier to work with RNS superstring formalism, rather than with bosonic string theory. The appearance of intrinsic AdS isometries and higher-spin currents is closely related with geometrical properties of the supermoduli space in RNS superstring theory.

So in this chapter, we shall concentrate on superstring amplitudes in the Neveu-Schwarz-Ramond (NSR) formalism, as the first step towards understanding how higher spin symmetries emerge in the full (unbroken) superstring theory. The NSR superstring theory possesses the local supersymmetry on the worldsheet (as a fermionic counterpart of the reparametrizations) but no manifest supersymmetry in space-time. Despite the fact that the NSR (Neveu-Schwarz-Ramond) superstring theory appears to be the best explored, elegant and relatively simple model describing the superstring dynamics, it still has a number of serious complications. One, and perhaps the most important of them is related to the problem of picture-changing and the difficulties in constructing the generating functional for Ramond and Ramond-Ramond states, because of the picture-changing ambiguity [63]. For example, since the Ramond vertex operators have fractional ghost-numbers, one always has to consider their combinations at different ghost pictures in order to calculate their scattering amplitudes (except for the 4-point function where all the operators can be taken at the picture $-1/2$). As a result, constructing the generating functional for the Ramond scattering amplitudes becomes a confusing question. Actually, the problem of the picture-changing is not limited to the Ramond sector. Due to the ghost number anomaly cancellation condition, vertex operators of any correlation function in NSR string theory must have total superconformal ghost number $-2$ and total $b-c$ ghost number 3. This means that we cannot limit ourselves to just integrated and picture 0 vertices in the generating functional of the NSR model, since both integrated and unintegrated vertex operators, at various superconformal pictures must be involved in the sigma-model action. Therefore, due to this picture ambiguity, we face the problem building the generating functional for the consistent perturbation theory for NSR strings.

Consider the string scattering amplitudes in the NSR formalism.To be certain, let us first consider the closed string case, the open strings can be treated similarly. The scattering amplitude on a sphere for N vertex operators in the NSR superstring theory is given by:

$$
\begin{aligned}
&< V_1(z_1,\bar{z}_1)...V_N(z_N,\bar{z_N}) > \\
&= \int \prod_{i=1}^{M(N)} dm_i d\bar{m}_i \int \prod_{a=1}^{P(N)} d\theta_a d\bar{\theta_a} \int DXD\psi D\bar{\psi}D[ghosts] \\
&e^{-S_{NSR}+m_i<\xi^i|T_m+T_{gh}>+\bar{m}_i<\bar{\xi}^i|\bar{T}_m+\bar{T}_{gh}>} \\
&\times e^{\theta_a<\chi^a|G_m+G_{gh}>+\bar{\theta}_a<\bar{\chi}_a|\bar{G}_m+\bar{G}_{gh}>} \\
&\prod_{a=1}^{M(N)} \delta(<\chi^a|\beta>)\delta(<\bar{\chi}^a|\bar{\beta}>) \\
&\times \prod_{i=1}^{P(N)} <\xi^i|b><\bar{\xi}^i|\bar{b}> V_1(z_1,\bar{z}_1)...V_N(z_N,\bar{z}_N)
\end{aligned}
\tag{2.1}
$$

Here $z_1, ..., z_N$ are the points of the vertex operator insertions on the sphere and

$$
\begin{aligned}
&S_{NSR} \sim \int d^2z\{\partial X_m\bar{\partial}X^m + \psi_m\bar{\partial}\psi^m + \bar{\psi}_m\partial\bar{\psi}^m + b\bar{\partial}c \\
&+\bar{b}\partial\bar{c} + \beta\bar{\partial}\gamma + \bar{\beta}\partial\bar{\gamma}\} \\
&m = 0, ..., 9
\end{aligned}
\tag{2.2}
$$

is the NSR superstring action in the superconformal gauge. Next, $(m_i, \theta_a)$ are the holomorphic even and odd coordinates in the moduli superspace and $(\xi^i, \chi^a)$ are their dual super Beltrami differentials (similarly for $(\bar{m}_i, \bar{\theta}_a)$ and $(\bar{\xi}^i, \bar{\chi}^a)$ The $< ...|... >$ symbol stands for the scalar product in the Hilbert space and the delta-functions $\delta(<\chi^a|\beta>)$ and $\delta(<\xi^i|b>) = <\xi^i|b>$ are needed to insure that the basis in the moduli space is normal to variations along the superconformal gauge slices (similarly for the antiholomorphic counterparts). The dimensions of the moduli and supermoduli spaces are related to the number N of the vertex operators present and are different for the NS and Ramond sectors. The standard bosonization relations for the $b, c, \beta$ and $\gamma$ ghosts are given by

$$\begin{aligned}
&c(z) = e^{\sigma}(z);\\
&b(z) = e^{-\sigma}(z)\\
&\gamma(z) = e^{\phi-\chi}(z); \beta(z) = e^{\chi-\phi}\partial\chi(z)\\
&< \sigma(z)\sigma(w) >=< \chi(z)\chi(w) >= - < \phi(z)\phi(w) >= log(z-w)
\end{aligned} \tag{2.3}$$

In NSR superstring theory, while the partition function on the sphere does not have any modular dependence, the integration over the moduli and the supermoduli does appear for the sphere scattering amplitudes, for all the N-point correlators with $N \geq 3$. This is related to the important fact that in string theory the BRST and the local gauge invariances are not necessarily equivalent. Namely, the physical vertex operators while being BRST-invariant, are not necessarily invariant under the local supersymmetry. Moreover, while the integrated vertices are invariant under the local reparametrizations, the unintegrated ones are not. E.g. the infinitezimal conformal transformation of an unintegrated photon at zero momentum gives:

$$\begin{aligned}
&\delta_\epsilon V_{ph}(z) = \oint \frac{dw}{2i\pi}\epsilon(w)T(w)(c\partial X^m + \gamma\psi^m)(z)\\
&= \oint \frac{dw}{2i\pi}\epsilon(w)\{\frac{1}{z-w}\partial(c\partial X^m + \gamma\psi^m)(z) + O((z-w)^0)\}\\
&= \epsilon(z)\partial(c\partial X^m + \gamma\psi^m)(z)
\end{aligned} \tag{2.4}$$

More generally, under the infinitezimal conformal transformations $z \to z + \epsilon(z)$ any dimension 0 primary field transforms as

$$\Phi(z) \to \Phi(z + \epsilon(z)) = \Phi(z) + \epsilon\partial\Phi(z)$$

Similarly, applying the local supersymmetry generator $\oint \frac{dw}{2i\pi}\epsilon(w)G(w)$ to a photon at picture $-1$: $V_{ph} = ce^{-\phi}\psi^m$ with G being a full matter+ghost supercurrent and $\epsilon$ now a fermionic parameter, it is easy to calculate

$$\delta_\epsilon V_{ph} = \epsilon(z)(-\frac{1}{2}ce^{-\phi}\partial X^m + c\partial ce^{\chi}\partial\chi e^{-2\phi}\psi^m) \neq 0.$$

Note, however, that the photon vertex operator at picture 0 is supersymmetric. The problem is that, due to the $b-c$ and $\beta-\gamma$ ghost number anomalies on the sphere we cannot limit ourself to just integrated vertices or those at the picture 0 but must always consider a combination of pictures in the amplitudes. This combination, however, is not arbitrary but must follow from the consistent superstring perturbation theory, or the expansion in the appropriate sigma-model terms. These terms are typically of the form $\lambda V$ where V is a vertex operator of some physical state and $\lambda$ is the space-time field corresponding to this vertex operator.This term must be added to the original action as a string emits the $\lambda$-field. Then the string partition function must be expanded in $\lambda$ leading to the perturbation theory series and the effective action for $\lambda$, upon computing the correlators. But the important question is - which vertex operator should we take? Should it be integrated or unintegrated, which superconformal ghost picture should be chosen? If we choose the integrated and picture 0 vertices as the sigma-model terms, the local superconformal symmetry is preserved, but we face the problem of the ghost number anomaly cancellation. For this reason these vertex operators are not suitable as the sigma-model terms. If, on the other hand, we choose unintegrated vertices at the picture -1, then one faces the question of the gauge non-invariance - though these operators are BRST-invariant, they are not invariant under local superconformal gauge transformations. Therefore, to ensure the gauge invariance of the scattering amplitudes, it is necessary to impose some restrictions on the parameter $\epsilon(z)$ of the superconformal transformations; namely, it must vanish at all the insertion points of the vertex operators. With such a restriction on the gauge parameter (effectively reducing the superconformal gauge group) the scattering amplitudes will be gauge invariant. Below we will prove that the consistent superstring perturbation theory must be defined as follows:

1) All the vertex operators must be taken unintegrated and at the picture $-1$.

2) The restrictions on the gauge parameter:

$$\epsilon(z_k) = 0; k = 1, ...N \tag{2.5}$$

must be imposed at the insertion points of the vertex operators (where $\epsilon$ is either a bosonic infinitezimal reparametrization or a fermionic local supersymmetry transformation on the worldsheet).

The restrictions (2.5) on the gauge parameter effectively reduce the gauge group of superdiffeomorphisms on the worldsheet. As a result, it is in general no longer possible to choose the superconformal gauge, fully eliminating the functional integrals over the worldsheet metric $\gamma^{ab}$ and the worldsheet gravitino field $\chi_\alpha^a$ in the partition function. Instead these functional integrals are reduced to integration over the finite number of the conformal moduli $m_i; i = 1, ..., M(N)$ and the anticommuting moduli $\theta^a; a = 1, ..., P(N)$ of the gravitino while the gauge-fixed superstring action is modified by the supermoduli terms. The final remark about the general expression for the correlation function concerns the dimensionalities of the odd and even moduli spaces as functions of the number N of the vertices. These dimensionalities are equal to the numbers of independent holomorphic super Beltrami differentials $\chi^a$ and $\xi^i$, dual to the basic vectors $\theta_a(z)$ and $m_i(z)$ in the supermoduli space. These numbers are different for the NS and Ramond sectors. Namely, the OPE's of the stress-tensor and the supercurrent with the unintegrated perturbtive vertex operators (e.g. a photon) are given by:

$$\begin{aligned} T(z)V(w) &\sim \frac{1}{z-w}\partial V(w) + O(z-w)^0 \\ G(z)V(w) &\sim \frac{1}{z-w}W(w) + O(z-w)^0 \end{aligned} \tag{2.6}$$

where W is some normally ordered operator of conformal dimension 1/2. E.g. for a picture zero photon at zero momentum: $V = c\partial X^m + \gamma\psi^m$ it is easy to check that $W(w) = \frac{1}{2}\partial c\psi^m$. Now since the induced worldsheet metric and the induced worldsheet gravitino field can be expressed as

$$\begin{aligned} &\gamma_{ab}(z) =: \partial_a X^m \partial_b X_m : (z) + ... \sim T_{ab}(z) \\ &\chi_{a\alpha}(z) =: \psi_\alpha^m \partial_a X^m : (z) + ... \sim G_{a\alpha}(z) \\ &a, \alpha = 1, 2 \end{aligned} \tag{2.7}$$

It is clear that the $\chi_{a\alpha}(z)$, $\gamma_{a\alpha}(z)$ and hence the corresponding superconformal moduli $\theta_a(z)$ and $m_i(z)$ behave as $(z-z_i)^{-1}; i = 1, ...N$ when approaching a NS vertex point. For this reason the natural choice of the basis for $m$ and $\theta$, consistent with their pole structures and the holomorphic properties, is given by

$$\begin{aligned} &\theta_a(z) = z^a \prod_{j=1}^{N} (z-z_j)^{-1} \\ &m_i(z) = z^i \prod_{j=1}^{N} (z-z_j)^{-1} \\ &a = 1, ..., M(N); i = 1, ..., P(N) \end{aligned} \tag{2.8}$$

The holomorphy condition requires that at the infinity the $\theta^a$ vectors must go to zero not slower than the supercurrent's two-point function $lim_{z\to\infty} < G(0)G(z) >\sim \frac{1}{z^3}$ while the $m^i(z)$ must decay as $lim_{z\to\infty} < T(0)T(z) >\sim \frac{1}{z^4}$ or faster. To see this, note that the expansions of T and G in terms of their normal modes are

$$T(z) = \sum_n \frac{L_n}{z^{n+2}}$$
$$G(z) = \sum_n \frac{G_n}{z^{n+3/2}} \tag{2.9}$$

therefore the normal orgering of these operators at zero point implies $n \leq -2$ for T and $n \leq -\frac{3}{2}$ for G. Under conformal transformation $z \to u = \frac{1}{z}$ mapping the zero point to infinity, G and T transform as

$$G(z) \to -iu^3 G(u)$$
$$T(z) \to u^4 T(u) + ... \tag{2.10}$$

therefore the normal ordering or the regularity at the infinity requires

$$T(u) \sim \frac{1}{u^4} + O(\frac{1}{u^5})$$
$$G(u) \sim \frac{1}{u^3} + O(\frac{1}{u^4}) \tag{2.11}$$

implying the same asymptotic behaviour for $\theta_a$ and $m_i$, in the light of (7). This condition immediately implies that

$$M(N) = N - 3$$
$$P(N) = N - 2 \tag{2.12}$$

for the scattering amplitudes in the NS-sector. For any N unintegrated Ramond operators $V_R$ the OPE of T and V remains the same as in (6), therefore the basis and and the number of the bosonic moduli do not change in the Ramond sector. However, the OPE of GSO-projected Ramond vertex operators with the supercurrent is given by

$$G(z)V_R(w) \sim (z-w)^{-\frac{1}{2}} U(w) + ... \tag{2.13}$$

where U(w) is some dimension 1 operator. For this reason the supermoduli of the gravitini, approaching the insertion points of $V_R(z_i)$, behave as $\chi^a(z) \sim (z - z_i)^{-\frac{1}{2}}$. For the scattering amplitudes on the sphere involving the $N_1$ NS vertices $V(z_i), i = 1, ...N_1$ and $N_2$ Ramond operators

$V(w_j), j = 1, ..., N_2$ the basis in the moduli space (involving the combinations of quadratic and 3/2-differentials) should be chosen as

$$\theta^a(z) = z^a \prod_{i=1,j=1}^{N_1,N_2} \frac{1}{(z - z_i)\sqrt{z - w_j}} \tag{2.14}$$

The holomorphy condition for these differentials gives the total dimension of the supermoduli space for the NSR N-point scattering amplitudes:

$$M(N) = N - 3$$
$$P(N) = N_1 + \frac{N_2}{2} - 2; N_1 + N_2 = N \tag{2.15}$$

This concludes the explanation of the formula for the NSR scattering amplitudes.

Performing the integration over $m_i$ and $\theta_a$, we obtain

$$\begin{aligned}
&< V_1^{NS-NS}(z_1, \bar{z}_1)...V_{N_1}^{NS-NS}(z_{N_1}, \bar{z}_{N_1}) \\
&V_1^{RR}(w_1, \bar{w}_1)...V_{N_2}^{RR}(w_{N_2}, \bar{w}_{N_2}) > \\
&= \int DXD\psi D\bar{\psi}D[ghosts]e^{-S_{NSR}} \\
&\prod_{a=1}^{N_1+\frac{1}{2}N_2-2} |\delta(<\chi^a|\beta>) < \chi^a|G_m + G_{gh} > |^2 \\
&\times \prod_{i=1}^{N_1+N_2-3} | < \xi^i|b > \delta(<\xi^i|T>)|^2 \\
&V_1^{NS-NS}(z_1, \bar{z}_1)...V_{N_1}^{NS-NS}(z_{N_1}, \bar{z}_{N_1}) \\
&V_1^{RR}(w_1, \bar{w}_1)...V_{N_2}^{RR}(w_{N_2}, \bar{w}_{N_2})
\end{aligned} \tag{2.16}$$

and similarly for the open string case. Each of the operators $\delta(<\chi^a|\beta>)$ $<\chi^a|G_m + G_{gh}>$ has the superconformal ghost number +1. These operators are the standard operators of picture-changing. Indeed, for the particular choice of $\chi^a = \delta(z - z_a)$ we have

$$\begin{aligned}
&: \Gamma : (z_a) = \delta(<\chi^a|\beta>) < \chi^a|G_m + G_{gh}> \\
&=: \delta(\beta)(G_m + G_{ghost})(z_a) =: e^{\phi}(G_m + G_{ghost}) : (z_a)
\end{aligned} \tag{2.17}$$

i.e. the standard expression for the picture-changing operator and similarly for $\bar{\Gamma}(\bar{z}_a)$. The total ghost number of the left or right picture changing insertions following from the supermoduli integration is therefore equal to $N_1 + \frac{1}{2}N_2 - 2$. Thus if one takes all the NS-NS vertex operators at the canonical $(-1,-1)$-picture and all the RR operators at picture $(-\frac{1}{2},-\frac{1}{2})$, the integration over the supermoduli insures the correct ghost number of the correlation function to cancel the ghost number anomaly on the sphere. Similarly, the Z-operator

$$Z =:< \xi^i|b > \delta(< \xi^i|T >) \tag{2.18}$$

resulting from the integration over the $m_i$-moduli has left and right fermionic ghost numbers $-1$.

Since we found that the total number of these operators is equal to $N-3$, one has to take all the vertex operators unintegrated (i.e. with the left and right $+1$ $b-c$ fermionic ghost number) so that the fermionic ghost number anomaly, equal to $-3$ on the sphere, is precisely cancelled by the moduli integration. The $Z$-operator is a straightforward generalization of the picture-changing transformation by $\Gamma$-operator for the case of the fermionic $b-c$ pictures. In particular, the integrated and the unintegrated vertex operators in NSR string theory are simply two different $b-c$ ghost picture representations of the vertex operator. The Z-operator must therefore reduce the $b-c$ ghost number of a vertex operator by one unit, in particular, transforming the local operators into the non-local , i.e. the dimension 0 ghost number 1 primaries into ghost number 0 dimension 1 primaries integrated over the worldsheet boundary dimension (1,1) operators integrated over the worldsheet). Namely, if $cV(z)$ is an unintegrated vertex of a conformal dimension zero, the $Z$-transformation should give

$$: Z(cV) :\sim \oint dzV(z) + ... \tag{2.19}$$

i.e. the $c\bar{c}V$-operator is transformed into the the worldsheet integral of $V$ plus possibly some other terms ensuring the BRST-invariance. Similarly, in closed string theory one has

$$Z =: b\bar{b}\delta(T)\delta(\bar{T}) :$$
$$: Z(c\bar{c})V^{(1,1)} := \int d^2zV^{(1,1)}(z,\bar{z}) \tag{2.20}$$

The non-locality of the $Z$-operator follows from the non-locality of the delta-function of the full stress-energy tensor, which has conformal dimension $-2$ but depends on fields with positive dimensions. As previously, one can choose the conformal coordinate patches on the Riemann surface excluding $N-3$ points corresponding to the basis $\xi^i(z) = \delta(z-z_i)$, so the $Z$-operator becomes

$$Z(z_i) =: b\delta(T) : (z_i) \tag{2.21}$$

This expression for the $Z$-operator is still not quite convenient for practical calculations. Using the BRST invariance of $Z$ its OPE properties we shall try to derive an equivalent representation for $:b\delta(T):$ (with the help of the arguments similar to those one uses to derive the exponential representations for $\delta(\gamma) = e^{-\phi}$ and $\delta(\beta) = e^{\phi}$). Since the $Z$ operator must have the conformal dimension 0, the operator $\delta(T)$ has conformal dimension $-2$. As the full matter+ghost stress tensor satisfies

$$T(z)T(w) \sim 2(z-w)^{-2}T(w) + (z-w)^{-1}\partial T(w) + :TT: + ..., \tag{2.22}$$

the corresponding OPE for $\delta(T)$ must be given by

$$:(z-w)^2\delta(T(z))\delta(T(w)) :\sim \delta(T(w)) + ... \tag{2.23}$$

In addition, the $\delta(T)$-operator must satisfy

$$[Q_{brst}, Z] =: T\delta(T) := 0 \tag{2.24}$$

since $Z$ is BRST-closed. This is altogether sufficient to fix the form of $Z$ in RNS theory. One finds

$$Z =: b\delta(T(w)) := \oint \frac{dz}{2i\pi}(z-w)^3\{bT - 4ce^{2\chi-2\phi}T^2 :\} \tag{2.25}$$

for open strings and

$$\begin{aligned}|Z|^2(w,\bar{w}) = \int d^2z|z-w|^6\{(:bT: - :4ce^{2\chi-2\phi}T^2:)\\ \times(\bar{b}\bar{T} - 4\bar{c}e^{2\bar{\chi}-2\bar{\phi}}\bar{T}^2:)\}\end{aligned} \tag{2.26}$$

for closed strings. Note that, just as the $\Gamma$-operator $\Gamma = \{Q_{brst}, \xi\}$ can be cast as a BRST commutator in the large Hilbert space, so is the case with the $Z$-operator:

$$Z = 16\{Q_{brst}, \oint \frac{dz}{2i\pi}(z-w)^3 bc\partial\xi\xi e^{-2\phi}T\} \tag{2.27}$$

By simple calculation, using $\{Q_{brst}, b\} = T$ and $\{Q_{brst}, ce^{2\chi-2\phi}\} = \frac{1}{4} - \partial cce^{2\chi-2\chi}$ it is easy to check that the integral representation (2.27) for the $b-c$ picture-changing operator is BRST-invariant. particularly maps the unintegrated vertices into integrated ones. The rules of how $Z(w)$ acts on unintegrated vertices are as follows. Let $cV(w)$ be an unintegrated vertex operator at $w$, where $V$ is the dimension 1 operator. Writing

$$Z(w) \equiv \int \frac{dz}{2i\pi}(z-w)^3 R(z) \tag{2.28}$$

where $R$ is the operator in the integrand in (2.27), one has to calculate the OPE between $R(z)$ and $cV(w)$ around the $z$-point. For elementary perturbative vertices, such as a graviton, the relevant contributions (up to total derivatives and integrations by parts) will be of the order of $(z-w)^{-3}$, cancelling the factor of $(z-w)^3$ in (2.27) and removing any dependence on $w$. The operator $W(z)$ from the $(z-w)^{-3}$ of the OPE of the conformal dimension 1 will then be the integrand of the vertex in the integrated form. As for the possible more singular terms of the OPE, one can show that for the perturbative superstring vertices, such as a graviton, are generally the total derivatives; higher order terms will be either total derivatives or BRST trivial (being of the general form $TL$ where L is BRST-closed).

Analogously, the operator of the $Z$-transformations in the closed string case is given by

$$Z_{closed}(w,\bar{w}) = \int d^2z|z-w|^6|R|^2(z,\bar{z}) \tag{2.29}$$

In the latter case, the Z-transformation, applied to the unintegrated closed string vertices $c\bar{c}V^{(1,1)}(w,\bar{w})$ produces the ghost number 0 closed string operators, integrated over the worldsheet, with the integrand given by the term in the OPE around $z$ proportional to $\sim |z-w|^6$.

As a concrete illustration, let us consider the $Z$-transformation of the picture zero unintegrated graviton. For simplicity and brevity, let us consider the graviton at zero momentum. The $k \neq 0$ case can be treated analogously, even though the calculations would be a bit more cumbersome. The expression for the unintegrated vertex operator of the graviton is given by

$$V(w)\bar{V}(\bar{w}) \sim (c\partial X^m + \gamma\psi^m)(\bar{c}\bar{\partial}X^n + \bar{\gamma}\bar{\psi}^n)$$

Consider the expansion of $R(z)$ with $V$ around the $z$-point (the OPE of $\bar{R}$ with $\bar{V}$ can be evaluated similarly). Let's start with the Z-transformation or the $c\partial X^m$ part of $V$. Consider the $bT$-term of $R(z)$ first. A simple calculation gives

$$\begin{aligned} &:bT:(z):c\partial X^m:(w) \sim (z-w)^{-2}(\partial^2 X^m(z) + :\partial\sigma\partial X^m:(z)) \\ &+O((z-w)^{-1}) \end{aligned} \tag{2.30}$$

Next, consider the OPE of $V$ with the remaining two terms of $R(z)$ (2.27).

The OPE calculation gives

$$:4ce^{2\chi-2\phi}TT - 4bc\partial ce^{2\chi-2\phi}:(z)c\partial X^m(w)$$
$$\sim 4(z-w)^{-2}:\partial\sigma c\partial ce^{2\chi-2\phi}\partial X^m:(z)$$
$$+O((z-w)^{-1}) \tag{2.31}$$

It is easy to check that the operator of the $(z-w)^{-2}$ term of this OPE can be represented as

$$:\partial\sigma c\partial ce^{2\chi-2\phi}\partial X^m:(z)$$
$$= -\frac{1}{4}\partial\sigma\partial X^m(z) + \{Q_{brst},:ce^{2\chi-2\phi}\}\partial\sigma\partial X^m:(z) \tag{2.32}$$

Next, since $\partial\sigma(z) = -:bc:(z)$ and since $\{Q_{brst}, c\partial X^m\} = 0$ due to the BRST invariance of the integrated vertex, we have

$$[Q_{brst}, \partial\sigma\partial X^m] = -[Q_{brst}, bc\partial X^m] = -cT\partial X^m, \tag{2.33}$$

therefore

$$:\{Q_{brst}, ce^{2\chi-2\phi}\}\partial\sigma\partial X^m:(z) = \{Q_{brst}, ce^{2\chi-2\chi}\partial\sigma\partial X^m\} \tag{2.34}$$

since

$$:ce^{2\chi-2\phi}[Q_{brst}, \partial\sigma\partial X^m]:=:cce^{2\chi-2\phi}T\partial X^m:=0 \tag{2.35}$$

as $:cc:=0$. For this reason

$$4\partial\sigma c\partial ce^{2\chi-2\phi}\partial X^m = -\partial\sigma\partial X^m + [Q_{brst},...] \tag{2.36}$$

and therefore the $(z-w)^{-2}$ order OPE term of (2.31) precisely cancels the corresponding $\partial\sigma\partial X^m$-term of the operator product (2.30) of $bT$ with $c\partial X^m$, up to the BRST-trivial piece. With some more effort, one similarly can show that the $(z-w)^{-1}$ and other higher order terms of the OPE of $R(z)$ with $c\partial X^m$ are the exact BRST-commutators. Proceeding similarly with the antiholomorphic OPE's, we obtain

$$:Zc\bar{c}\partial X^m\bar{\partial}X^n:(w,\bar{w})$$
$$= \int d^2z|z-w|^2\partial^2X^m\bar{\partial}^2X^n(z,\bar{z}) + [Q_{brst}...] \tag{2.37}$$

Finally, integrating twice by parts we get

$$:Zc\bar{c}\partial X^m\bar{\partial}X^n:= \int d^2z\partial X^m\bar{\partial}X^n(z,\bar{z}) + [Q_{brst},...] \tag{2.38}$$

Next, consider the $\gamma\psi^m$-part of the graviton's unintegrated vertex (left and right-moving alike). The calculation gives

$$:R:(z)\gamma\psi^m(w) \sim (z-w)^{-3}\partial(ce^{\chi-\phi}\psi^m)$$
$$+(z-w)^{-2}[Q_{brst},:bc\beta\psi^m:] + [Q_{brst},...] \tag{2.39}$$

Performing the analogous calculation for the right-moving part and getting rid of the total derivatives we obtain

$$: Z\gamma\bar{\gamma}\psi^m\bar{\psi}^n := [Q_{brst}, ...] \tag{2.40}$$

Thus the full $Z$-transformation of the unintegrated graviton gives

$$: ZV_{grav}^{unintegr} := \int d^2z\partial X^m\bar{\partial}X^n(z,\bar{z}) + [Q_{brst}, ...] \tag{2.41}$$

Thus the result is given by the standard integrated vertex operator of the graviton, up to BRST-trivial terms. Note that if one naively applies the picture-changing operator : $\Gamma$ : to the integrated photon $V_{ph}^{(-1)} = \oint \frac{dz}{2i\pi}e^{-\phi}\psi^m$, one gets $\oint \frac{dz}{2i\pi}(\partial X^m + c\beta\psi^m)$. The last term in this expression is not BRST-invariant, therefore at the first glance the picture-changing operation seems to violate the BRST-invariance for some integrated vertices. This contradiction is due to the fact that one is not allowed to straightforwardly apply the $\Gamma$-operator (which is the expression for the picture-changing consistent with the supermoduli integration) to the integrated vertices because, roughly speaking, the $\Gamma$-operation does not "commute" with the worldsheet integration. More precisely, the contradiction arises because the "naive" application of the $\Gamma$ picture-changing operator to the integrands ignores the OPE singularities between $\Gamma$ and $Z$ (the latter can be understood as the "operator of the worldsheet integration"). On the other hand, if one uses the "conventional" definition of the picture-changing operation given by $[Q_{brst}, e^{\chi}V]$ then one gets

$$\{Q_{brst}, \oint \frac{dz}{2i\pi}e^{\chi-\phi}\psi^m\} = \oint \frac{dz}{2i\pi}(\partial X^m + \partial(ce^{\chi-\phi}\psi^m)) \tag{2.42}$$

The second total derivative in this expression can be thrown out and we get the picture 0 photon. Now it is clear that the definition of the picture-changing as $[Q_{brst}, e^{\chi}V]$ is consistent with the supermoduli integration and the expression for the local picture-changing operator, only if one accurately accounts for the the singularities of the OPE between $\Gamma$ and $Z$.

We have shown that the $Z$-transformation of the unintegrated graviton reproduces the full picture zero expression for the integrated vertex operator, up to BRST-trivial terms. Therefore the consistent procedure of the picture-changing implies that one always applies the picture-changing operator $\Gamma$ to unintegrated vertices, with the subsequent $Z$-transformation, if necessary, i.e. the $\beta\gamma$ picture-changing must be followed by the $Z$-transformation and not otherwise. Another useful expression is the Z-transformation of the picture-changing operator, or the integrated form of the picture-changing. Applying the $Z$-operator to $:\Gamma\bar{\Gamma} :=: e^{\phi+\bar{\phi}}G\bar{G}:$ one obtains

$$(\Gamma\bar{\Gamma})_{int}(w) =: Z\Gamma : (w)$$
$$= \int d^2z|z-w|^2\{: b\Gamma : -4 : ce^{2\chi-2\phi}(T - b\partial c)\Gamma :\} \times c.c.$$
$$\equiv \int d^2z|z-w|^2 P(z)\bar{P}(\bar{z}) \qquad (2.43)$$

Similarly, the expression for the single left or right integrated $\Gamma_{int}$ can be written as

$$\Gamma_{int} = \oint \frac{dz}{2i\pi}(z-w)P(z) \qquad (2.44)$$

The main advantage of using $:\Gamma_{int}:$ is the absence of singularities in the OPE between $\Gamma_{int}(w_1)\Gamma_{int}(w_2)$ which can be checked straightforwardly. The singularities in the $\Gamma\Gamma$ operator products of the usual (unintegrated) picture-changing operators are well-known to result in complications and inconsistencies in the picture-changing procedure. As we saw, these complications and the appearance of the singularities are due to the fact that, strictly speaking, the picture-changing procedure is not well-defined without the appropriate $Z$-transformations.

## 2.2 Ghost-Matter Mixing and Moduli Space Singularities

The scattering amplitude (2.1) involves the insertion of picture-changing operators for bosonic and fermionic ghosts, which precise form depends on the choice of basis for super Beltrami quadratic and 3/2−differentials. It has been shown [57] that the scattering amplitudes are invariant under the small variations of the Beltrami basis, up to the total derivatives in the moduli space. In particular, if one chooses the delta-functional basis for $\xi^i$ and $\chi^a$, this symmetry implies the independence on the insertion points of picture-changing operators. The situation is more subtle, however, when the picture-changing insertions $z_a$ of coincide with locations of the vertex operators (which precisely is the case for the amplitudes involving combinations of the vertex operators at different pictures). The equations (2.7), (2.8) and (2.11) imply the singular behavior of the supermoduli approaching the locations of the vertex operators. Namely, by simple conformal transformations it is easy to check that the singularities of (2.7), (2.8) and (2.11) at $z_i$ and $w_j$ correspond to orbifold points in the moduli space. As it has been pointed out in [57], if picture-changing operators are located at the orbifold points of the moduli space, the picture-changing gauge symmetry is reduced to the discrete automorphism group corresponding to all the possible permutations of the p.c. operators between these orbifold points. In particular, it's easy to see that for $N_1$ Neveu-Schwarz and $N_2$ Ramond perturbative vertex operators having a a total superconformal ghost number g, the volume of this automorphism group is given by

$$\Xi_{N_1,N_2}(g) = (N_1 + N_2)^{N_1+\frac{N_2}{2}+g} \tag{2.45}$$

The appearance of this discrete group is particularly a consequence of the polynomial property of picture-changing operators:

$$:\Gamma^m :: \Gamma :^n \sim: \Gamma^{m+n} : +[Q_{BRST}, ...] \tag{2.46}$$

which holds as long as the picture-changing transformations are applied to the perturbative string vertices, such as a graviton or a photon, which are equivalent at all the ghost pictures. However, apart from the usual massless states such as a graviton or a photon, the spectra of open and closed NSR strings also contain BRST-invariant and non-trivial vertex operators which cannot be interpreted in terms of emissions of point-like particles by a string. In case of an open string, an example of such a vertex operator is an antisymmetric 5-form, given by

$$\begin{aligned} V_5^{open}(k) &= H_{m_1...m_5}(k) c e^{-3\phi} \psi_{m_1}...\psi_{m_5} e^{ikX} \\ V_{5int}^{open}(k) &= H_{m_1...m_5}(k) \oint \frac{dz}{2i\pi} e^{-3\phi} \psi_{m_1}...\psi_{m_5} e^{ikX} + [Q_{brst}, ...] \end{aligned} \tag{2.47}$$

This vertex operator is physical, i.e. BRST-invariant and non-trivial, provided that the appropriate constraints are imposed on $H$ space-time field. In particular, BRST non-triviality of the operator requires that the $H$ five-form is not closed:

$$k_{[m_1} H_{m_2...m_6]} \neq 0 \tag{2.48}$$

The vertex operator (2.47) exists only at nonzero ghost pictures below $-3$ and above $+1$ , i.e. its coupling with the ghosts is more than just an artefact of a gauge and cannot be removed by picture-changing transformations. This situation is referred to as the ghost-matter mixing. In the closed string sector, an important example of the ghost-matter mixing vertex operator can be obtained by multiplying the five-form (2.47) by antiholomorphic photonic part:

$$V_5^{(-3)} = H_{m_1...m_5m_6}(k)c\bar{c}e^{-3\phi-\bar{\phi}}\psi_{m_1}...\psi_{m_5}\bar{\psi}_{m_6}e^{ikX}$$
$$V_{5int}^{(-3)} = H_{m_1...m_5m_6}(k)\int d^2ze^{-3\phi-\bar{\phi}}\psi_{m_1}...\psi_{m_5}\bar{\psi}_{m_6}e^{ikX}(z,\bar{z})$$
$$+[Q_{brst},...] \qquad (2.49)$$

where the 6-tensor $H_{m_1...m_6}$ is antisymmetric in the first five indices. The BRST-invariance and non-triviality conditions for this vertex operator imply

$$k_{[m_7}H_{m_1...m_5]m_6}(k)\neq 0$$
$$k_{m_6}H_{m_1...m_5m_6}(k) = 0 \qquad (2.50)$$

These constraints particularly entail the gauge transformations for the H-tensor

$$H_{m_1...m_5m_6}(k) \to H_{m_1...m_5m_6}(k) + k_{[m_1}R_{m_2...m_5]m_6}(k) \qquad (2.51)$$

where $R$ is a rank 5 tensor antisymmetric over the first 4 indices, satisfying

$$k_{m_6}R_{m_2...m_6} = 0$$

It is easy to check that the BRST constraints (2.50) including the related gauge transformations (2.51) eliminate 1260 out of 2520 independent components of the H-tensor. Therefore the total number of the degrees of freedom related to the closed string vertex operator for the $H$-field is equal to 1260. Their physical meaning can be understood if we note that the BRST conditions (2.50) imply that for each particular polarization $m_1...m_6$ of the vertex operator (2.49) the momentum $k$ must be normal to the directions of the polarization, i.e. confined to the four-dimensional subspace orthogonal to the $m_1,...m_6$ directions. The number of independent polarizations of $V_5$ in 10 dimensions is equal to $\frac{10!}{4!6!} = 210$ therefore the total number of degrees of freedom per polarization is equal to 6. For instance consider $m_1 = 4,...m_6 = 9$ so that $k$ is polarized along the $0,1,2,3$ directions.Then we can make a $4+6$-split of the space-time indices $m \to (a,t); a = 0,...3; t = 4,...,9; H_{m_1...m_6} \equiv H_{t_1...t_6}$ The tensor $H_{t_1...t_6}$ is antisymmetric in the first five indices Since the total number of independent degrees of freedom for this polarization is equal to 6, one can always choose it in the form $t_6 \neq t_i, i = 1,...,5$ by using suitable gauge transformations (2.51). This means in turn that one can choose the basis

$$\lambda_t = H_{t_1...t_5t},$$
$$t = 4,...9; t \neq t_1,...t_5 \qquad (2.52)$$

with $H$ antisymmetrized over $t_1, ..., t_5$. It is easy to see that the $\lambda_t$ simply parametrize the 6 physical degrees of freedom for this particular polarization of $V_5$. To understand the meaning of $\lambda_t$-field one has to calculate its effective action. Computing the closed string 4- point correlation function $< V_5(z_1, \bar{z_1})...V_5(z_4, \bar{z_4}) >$ and the three-point function $< V_5 V_5 V_\varphi >$ of two $V_5$'s with the dilaton one can obtain that they reproduce the appropriate expansion terms of the DBI effective action for the D3-brane

$$S_{eff}(\lambda) = \int d^4 x e^{-\varphi} \sqrt{det(\eta_{ab} + \partial_a \lambda_t \partial_b \lambda^t)} \tag{2.53}$$

where $x$ is a Fourier transform of $k$. The open string vertices (2.47) can also be shown to carry the Ramond-Ramond charges which can be demonstrated by calculating their disc correlation functions with the appropriate RR 5-form operator. That is, the disc correlation function $< V_5^{(-3)}(k) V_5^{(+1)}(p) V_{RR}^{(+1/2,-1/2)}(q) >$, in which the RR vertex operator has to be taken at the $(+1/2, -1/2)$-picture, is linear in the momentum leading to the term $\sim (dH)^2 A_{RR}$ in the effective action, where $A$ is the Ramond-Ramond 4-form potential. This implies that the $dH$-field corresponds to the wavefunction of the RR-charge carrier, i.e. of the D-brane. The BRST nontriviality condition (2.48) for the open string $V_5$-vertices simply means that this vawefunction does not vanish. Thus the closed string $V_5$-operators generate the kinetic term of the D-brane action while the open string $V_5$-vertices account for its coupling with the RR-fields. The fact that the closed-string amplitudes lead to the D-brane type dilaton coupling of the effective action is related to the non-perturbative nature of the $V_5$-vertices, which in turn is the consequence of the ghost-matter mixing, or the picture inequivalence of $V_5$. Let us explore this inequivalence in more details, from the point of view of the supermoduli geometry. The OPE of the full matter+ghost supercurrent $G(z) = -\frac{1}{2}\psi_m \partial X^m - \frac{1}{2} b\gamma + c\partial\beta + \frac{3}{2}\beta\partial c$ with $V_5(w)$ gives

$$G(z) V_5(w) \sim -\frac{1}{2(z-w)^3} H_{m_1...m_5} c\partial c e^{\chi - 4\phi} \psi_{m_1}...\psi_{m_5} e^{ikX} + ... \tag{2.54}$$

and similarly for the OPE of $G$ with the left-moving part of the closed string $V_5$. This means that the supermoduli of the gravitini behave as $\theta^a(z) \sim (z - z_k)^{-3}$ as they approach insertion points of $V_5$. Such a behavior of the supermoduli approaching the $V_5$ operators is much more singular than of those approaching the usual perturbative vertices. These singularities no longer correspond to the orbifold points of the moduli space. Instead they entirely overhaul the moduli space topology, effectively creating boundaries and global curvature singularities. To illustrate this consider the worldsheet metric

$$ds^2 = dzd\bar{z} + z^{-\alpha}(dz)^2 + \bar{z}^{-\alpha}(d\bar{z})^2 \tag{2.55}$$

In terms of the $r, \varphi$ coordinates where $z = re^{i\varphi}, \bar{z} = re^{-i\varphi}$ this metric is given by

$$\begin{aligned}&(1 + 2r^{-\alpha}cos(\alpha - 2)\varphi)dr^2 + (1 - 2r^{-\alpha}cos(\alpha - 2)\varphi)r^2d\varphi^2\\&-4r^{1-\alpha}sin(2 - \alpha)\varphi drd\varphi\end{aligned} \tag{2.56}$$

The area of the disc of radius $\epsilon$ surrounding the origin point is given by

$$A = \int_0^{\epsilon} dr \int_0^{2\pi} d\varphi\sqrt{\gamma} = \int_0^{2\pi} d\varphi \int_0^{\epsilon} drr\sqrt{1 - 2r^{-2\alpha}} \tag{2.57}$$

This integral is of the order of $\epsilon^2$ for positive $\alpha$, $\epsilon^{2-\alpha}$ for $0 \leq \alpha < 2$ (this value reflects the deficit of the angle near the orbifold points) but it diverges if $\alpha \geq 2$ which means that for $\alpha = 3$ the disc is no longer compact but the origin point is blown up to become a global singularity. For this reason the scattering amplitudes in the presence of the $V_5$-vertices are no longer invariant under the discrete automorphism group. This means that the OPE's involving the $V_5$-operators become picture-dependent. Therefore, in order to correctly describe the physical processes involving the $V_5$-operators one has to sum over all the previously equivalent gauges, i.e. all the admissible pictures (corresponding to the possible locations of the picture-changing operators at the singularity points of the moduli space) and normalize by the volume of the original gauge group of automorphisms. The presence of the $V_5$-operators also modifies the suitable choice of the basis for the supermoduli and the number of independent 3/2-differentials For the amplitudes involving the total number $N_1$ of the $V_5$-insertions and $N_2$ of the standard perturbative vertices (without the ghost-matter mixing) the basis for $\theta^a$ is given by

$$\theta_a(z) = z^a \prod_{i=1}^{N_1}(z - z_i)^{-3} \prod_{j=1}^{N_2}(z - w_j)^{-1} \tag{2.58}$$

Accordingly the holomorphy condition (2.58) implies that in this case the number of the 3/2-differentials (equal to the number of the picture-changing insertions) is equal to

$$P(N) = 3N_1 + N_2 - 2 \tag{2.59}$$

As previously, the supermoduli integration insures the correct total ghost number of the vertex operators in the amplitude provided that the perturbative vertices of the generating sigma-model are all taken at pictures $-1$ while the ghost-matter mixing $V_5$-vertices are at the picture $-3$.

When a string propagates in flat space-time emits particles or solitons, the background is perturbed by the appropriate vertex operators. These operators multiplied by the corresponding space-time fields must be added to the original NSR superstring action. The condition of conformal invariance then leads to the effective equations of motion for these space-time fields and to the corresponding effective field theory. We conclude this section by writing down the generating functional for the NSR string sigma-model, free of the picture-changing ambiguities. The partition function of superstring theory perturbed by the set $\{V_i\}$ of physical vertex operators is given by

$$Z(\varphi_i) = \int DXD\psi D\bar{\psi}D[ghosts]e^{-S_{NSR}+\varphi_i V^i}\rho_{(\Gamma;Z)}\rho_{(\bar{\Gamma};\bar{Z})} \quad (2.60)$$

where $\rho_{(\Gamma;Z)}$ is the picture-changing factor due to the appearance of the supermoduli when one expands in $\varphi_i$. Though this factor enters differently for each term of the expansion, it is straightforward to check that, due to the ghost number conservation, one can recast it in the invariant form (independent on the order of expansion)

$$\begin{aligned}\rho_{(\Gamma;Z)} &= \sum_{m,n=0}^{\infty} \Xi_\xi^{-1}(n)\Xi_\chi^{-1}(m)\\ &\times \sum_{\{\xi^{(1)},..\xi^{(n)},\chi^{(1)}...\chi^{(m)}\}} \delta(<\chi^{(1)}|\beta>)<\chi^{(1)}|G>...\\ &\delta(<\chi^{(m)}|\beta>)<\chi^{(m)}|G>\\ &<\xi^{(1)}|b>\delta(<\xi^{(1)}|T>)...<\xi^{(n)}|b>\delta(<\xi^{(n)}|T>)\end{aligned} \quad (2.61)$$

and accordingly for $\rho_{(\bar{\Gamma};\bar{Z})}$ where the sum over $\xi^{(i)}$ and $\chi^{(i)}$ implies the summation over all the basic vectors of the $(m,n)$-dimensional spaces of super Beltrami differentials. Note that due to the ghost number anomaly cancellation condition, for any N-point correlator appearing as a result of expansion in in $\lambda$ only the $m = N-3$ and $n = N_1+\frac{1}{2}N_2-2$ terms of $\rho_{(\Gamma;Z)}$ contribute to the sigma-model. $\Xi_\xi(m)$ and $\Xi_\chi(n)$ are the volumes of the symmetry groups related to the picture-changing gauge symmetry (defined separately for the left and the right picture-changing). In the important case when the basic vectors are chosen at the orbifold points of the moduli spaces, the volumes are given by the relation (2.45). It's easy to see that in the picture-independent case inserting the $\rho_{(\Gamma,Z)}$-factor in the partition function can be reduced to the trivial statement that if one sums over N equivalent amplitudes with ghost picture combinations of the vertices and then divides by N, one gets the value of the amplitude. In the ghost-matter mixing cases involving the global singularities in the moduli space, the situation is more complicated. Thus the generating functional of the NSR sigma-model is a straightforward consequence of the expression (2.16) for the scattering amplitudes derived from the supermoduli integration. As was already said above, the operators $V_i$ of the sigma-model action (2.60) are taken unintegrated at picture $-1$ (and those with the ghost-matter mixing are at the picture $-3$). These operators are generally BRST invariant but not invariant under superdiffeomorphisms which simply means that the gauge symmetry, related to global conformal transformations, is fixed from the very beginning in the string sigma-model. The choice of the insertion points of the vertex operators corresponds to the choice of the Koba-Nielsen's measure in the correlation functions. Indeed, the $Z$-operators appearing as a result of the integration over the bosonic moduli, transform $N-3$ out of $N$ vertex operators into the integrated ones, while the remaining 3 are left in the unintegrated form $\sim c\bar{c}V(z_i,\bar{z}_i); i=1,2,3.$. Then the $c$ and $\bar{c}$-fields contribute the factor of $\prod_{i,j}|z_i-z_j|^2$ which precisely is the invariant Koba-Nielsen's measure [71] necessary for the calculations of the string scattering amplitudes. This concludes our analysis of how the singularities in the superconformal moduli lead to inequivalence of the ghost pictures. This inequivalence is crucial for understanding of how higher-spin modes with unbroken higher-spin symmetries emerge in the "larger" string theory, which will be discussed in the next chapter.

# Chapter 3

# Higher Spin Dynamics in a Larger String Theory

## 3.1 Ghost Cohomologies and Hidden Space-Time Symmetries in String Theory

The main lesson that we learned from the previous chapter is that the presence of global singularities in the supermoduli space leads to the violation of the ghost pictures for physical operators in superstring theory. This leads to the concept of *ghost cohomologies*, that will play the crucial role in the emergence of the extra space-time symmetries. As we know, in string theory the physical states are described by BRST-invariant and nontrivial vertex operators, corresponding to various string excitations and are defined up to transformations by the picture-changing operator $\Gamma = \{Q_{brst}, \xi\}$ and its inverse $\Gamma^{-1} = -4c\partial\xi e^{-2\phi}$ where $\xi = e^{\chi}$ and $\phi, \chi$ is the pair of the bosonized superconformal ghosts: $\beta = \partial\xi e^{-\phi}$ and $\gamma = e^{\phi-\chi}$. Acting with $\Gamma$ or $\Gamma^{-1}$ changes the ghost number of the operator by 1 unit, therefore each perturbative string excitation (such as a photon or a graviton) can be described by an infinite set of physically equivalent operators differing by their ghost numbers, or the ghost pictures. Typically, for a picture n operator one has

$$:\Gamma V^{(n)}:= V^{(n+1)} + \{Q_{brst}, ...\}$$

and

$$:\Gamma^{-1} V^{(n)}:= V^{(n-1)} + \{Q_{brst}, ...\}$$

The inverse and direct picture-changing operators satisfy the OPE identity

$$\begin{aligned}
&\Gamma(z)\Gamma^{-1}(w) = 1 + \{Q_{brst}, \Lambda(z,w)\}\\
&\Lambda(z,w) =: \Gamma^{-1}(z)(\xi(w) - \xi(z)) :
\end{aligned} \tag{3.1}$$

The standard perturbative superstring vertices can thus be represented at any integer (positive or negative) ghost picture. Such a discrete picture changing symmetry is the consequence of the discrete automorphism symmetry in the space of the supermoduli, described in the previous chapter. Varying the location of picture-changing operator (or, equivalently, varying the super Beltrami basis) inside correlation functions changes them by the full derivative in the supermoduli space. This ensures their picture invariance after the appropriate moduli integration, if the supermoduli space has no boundaries or global singularities. As was pointed out above, the global singularities, however, do appear in case if a correlation function contains vertex operators $V(z_n)$ such that the supermoduli coordinates diverge faster than $(z - z_n)^{-2}$ when they approach the insertion points on the worldsheet. If the latter is the case, the moduli integration of the full derivative term would result in the nonzero boundary contribution and the correlation function would be picture-dependent. We already have seen the example of an operator violating the equivalence of pictures in the previous chapter. Another crucially important physical example of how picture-dependent operators emerge in superstring theory is the following. Consider NSR superstring theory (critical or noncritical) in $d$ dimensions. With the super Liouville part included, its wordlsheet action is given by

$$
\begin{aligned}
&S_{NSR} = \frac{1}{2\pi}\int d^2z\{-\frac{1}{2}\partial X_m\bar{\partial}X^m - \frac{1}{2}\psi^m\bar{\partial}\psi^m - \frac{1}{2}\bar{\psi}_m\partial\bar{\psi}^m \\
&+b\bar{\partial}c + \bar{b}\partial\bar{c} + \beta\bar{\partial}\gamma + \bar{\beta}\partial\bar{\gamma}\} + S_{Liouville} \\
&S_{Liouville} = \frac{1}{4\pi}\int d^2z\{\partial\varphi\bar{\partial}\varphi + \lambda\bar{\partial}\lambda + \bar{\lambda}\partial\bar{\lambda} - F^2 \\
&+2\mu_0 b e^{b\varphi}(ib\lambda\bar{\lambda} - F)\}
\end{aligned}
\tag{3.2}
$$

This action is obviously invariant under two global $d$-dimensional space time symmetries - Lorenz rotations and translations, with the generators given by (1.21). It does have, however, yet another surprising and nontrivial space-time symmetry, which, to our knowledge, has not been discussed in the literature so far. That is, one can straightforwardly check that the action (3.2) is invariant under the following non-linear global transformations, mixing the matter and the ghost sectors of the theory:

$$
\begin{aligned}
&\delta X^m = \epsilon\{\partial(e^{\phi}\psi^m) + 2e^{\phi}\partial\psi^m\} \\
&\delta\psi^m = \epsilon\{-e^{\phi}\partial^2 X^m - 2\partial(e^{\phi}\partial X^m)\} \\
&\delta\gamma = \epsilon e^{2\phi-\chi}(\psi_m\partial^2 X^m - 2\partial\psi_m\partial X^m) \\
&\delta\beta = \delta b = \delta c = 0
\end{aligned}
\tag{3.3}
$$

where $\epsilon$ is constant global bosonic parameter. The invariance is straightforward to check. Substituting the transformations into the $X$-part of the superstring action and integrating by parts, we have:

$$\delta S_X = \delta(-\frac{1}{4\pi}\int d^2z\partial X_m\bar{\partial}X^m) = \frac{1}{2\pi}\int d^2z\delta X_m\partial\bar{\partial}X^m$$
$$= \frac{\epsilon}{2\pi}\int d^2z\{\partial(e^\phi\psi_m) + 2e^\phi\partial\psi_m\}\partial\bar{\partial}X^m$$
$$= \frac{\epsilon}{2\pi}\int d^2z\{-e^\phi\psi_m\partial^2\bar{\partial}X^m + 2e^\phi\partial\psi_m\partial\bar{\partial}X^m\} \tag{3.4}$$

Next, the transformation of the holomorphic $\psi$-term of the worldsheet action under (3.2) is

$$\delta(-\frac{1}{4\pi}\int d^2z\psi_m\bar{\partial}\psi^m) = \frac{1}{2\pi}\int d^2z\bar{\partial}(\delta\psi_m)\psi^m$$
$$= -\frac{\epsilon}{2\pi}\int d^2z\bar{\partial}\{e^\phi\partial^2X_m + 2\partial(e^\phi\partial X_m)\}\psi^m$$
$$= -\frac{\epsilon}{2\pi}\int d^2z\{\bar{\partial}(e^\phi\partial^2X_m)\psi^m - 2\bar{\partial}(e^\phi\partial X_m)\partial\psi^m\} \tag{3.5}$$

Adding these contributions together, we get

$$\delta S_X + \delta S_\psi \equiv \delta(-\frac{1}{4\pi}\int d^2z\{\partial X_m\bar{\partial}X^m + \psi_m\bar{\partial}\psi^m\})$$
$$= -\frac{\epsilon}{2\pi}\int d^2z(\bar{\partial}e^\phi)\{\partial^2X_m\psi^m - 2\partial X_m\partial\psi^m\} \tag{3.6}$$

Finally, using the OPE relation

$$lim_{z\to w}:\beta:(z):e^{2\phi-\chi}:(w)$$
$$\equiv lim_{z\to w}:e^{\chi-\phi}\partial\chi:(z):e^{2\phi-\chi}(w):=-:e^\phi:(z) \tag{3.7}$$

and the fact that $\delta\beta = 0$ we find the transformation of the $\beta-\gamma$ term of $S_{NSR}$ to be given by

$$\delta S_{\beta\gamma} \equiv \delta(\frac{1}{2\pi}\int d^2z\beta\bar{\partial}\gamma) = -\frac{\epsilon}{2\pi}\int d^2z\bar{\partial}\beta\bar{(}\delta\gamma)$$
$$= \frac{\epsilon}{2\pi}\int d^2z(\bar{\partial}e^\phi)\{\partial^2X_m\psi^m - 2\partial X_m\partial\psi^m\}, \tag{3.8}$$

i.e.

$$\delta(S_X + S_\psi + S_{\beta\gamma}) = 0 \tag{3.9}$$

Since the $b-c$ part of the superstring action (3.2) is manifestly invariant under (3.3), we have $\delta S_{NSR} = 0$, so the transformations (3.3) constitute the new space-time symmetry in superstring theory. The generator of this symmetry is easily deduced to be given by

$$T^{+}=\oint\frac{dz}{2i\pi}e^{\phi}(\psi_m\partial^2X^m-2\partial\psi_m\partial X^m)(z)\tag{3.10}$$

The integrand of this generator is a primary field of conformal dimension 1 (the dimension of $e^\phi$ is $-\frac{3}{2}$ and the matter factor is a primary of dimension $\frac{5}{2}$). There also exists another, dual version of this generator at ghost picture $-3$ (note that conformal dimensions of operators $e^{n\phi}$ and $e^{-(n+2)\phi}$ are equal and given by $-\frac{n^2}{2}-n$). It is generated by

$$T^{-}=\oint\frac{dz}{2i\pi}e^{-3\phi}(\psi_m\partial^2X^m-2\partial\psi_m\partial X^m)(z)\tag{3.11}$$

with the corresponding space-time symmetry transformations given by

$$\begin{aligned}
\delta X^m &= \epsilon\{\partial(e^{-3\phi}\psi^m)+2e^{\phi}\partial\psi^m\}\\
\delta\psi^m &= \epsilon\{-e^{-3\phi}\partial^2X^m-2\partial(e^{\phi}\partial X^m)\}\\
\delta\beta &= \epsilon e^{-4\phi+\chi}\partial\chi(\frac{1}{2}B^{(2)}_{-3\phi}F_{\frac{5}{2}}+B^{(1)}_{-3\phi}\partial F_{\frac{5}{2}}+\frac{1}{2}\partial^2F_{\frac{5}{2}})\\
\delta\gamma &= \delta b=\delta c=0
\end{aligned}\tag{3.12}$$

where

$$F_{\frac{5}{2}}=\psi_m\partial^2X^m-2\partial\psi_m\partial X^m\tag{3.13}$$

and $B^{(n)}_{f(\phi)}$ are the degree $n$ Bell polynomials in the $z$-derivatives of $f(\phi(z))$ (see (6.48)-(6.49) for the precise definition). Several important remarks should be made about the space-time symmetry transformations (3.3), (3.12) and their generators. First of all, up to BRST-exact terms, the generators (3.10), (3.11) commute with the standard Poincare space-time symmetry transformations (translations and rotations). These space-time symmetries, realized nonlinearly, are completely disentangled from the standard Poincare algebra. Second, an important property of the generators (3.10), (3.11) is their superconformal ghost dependence, violating the equivalence of pictures. That is, the generator (3.11) exists at its canonical negative picture $-3$ and can be transformed to lower negative pictures $n<-3$ by using the inverse picture-changing operator $\Gamma^{-1}$. It is, however, annihilated by the direct picture-changing operator $\Gamma$ and thus does not exist at ghost pictures above $-3$. Thus its ghost coupling is essential (not the matter of the gauge choice) and can be classified by its canonical picture $-3$, determined by the asymptotic behavior of the supermoduli, that diverge cubically when approaching the insertion point of the $T^{(-)}$

integrand. The generator $T^{(-)}$ is BRST-invariant and BRST nontrivial. The operator $T^{(+1)}$ carries the ghost number +1 and has no analogues at picture 0 and below, with the supermoduli as well diverging Similarly, the generator $T^{(+)}$ exists at its canonical positive picture +1 with the supermoduli as well diverging cubically approaching its integrand. Although it is not BRST-invariant (it only commutes with the stress-energy terms of $Q_{brst}$ but not the supercurrent terms), its BRST-invariance can be restored by adding $b-c$ ghost dependent terms, as will be explained below. The full BRST-invariant operator $T^{(+)}$ can be transformed by direct picture-changing operator $\Gamma$ to positive pictures 2 and above; it is annihilated by the inverse picture-changing transformation $\Gamma^{-1}$ at its minimal positive picture +1. Thus the space-time symmetry transformations (3.3), (3.12) can be classified by their minimal positive and negative ghost pictures +1 and −3. Are there any higher ghost picture analogues of such space-time symmetries and what is their geometrical significance? Before elaborating on that, it is useful to give a formal definition of ghost cohomologies, crucial for what follows.

1) The positive ghost number $N$ cohomology $H_N (N>0)$ is the set of physical (BRST invariant and non-trivial) vertex operators that exist at positive superconformal ghost pictures $n \geq N$ and that are annihilated by the inverse picture-changing operator $\Gamma^{-1} = -4c\partial\xi e^{-2\phi}$ at the picture $N$. $:\Gamma^{-1}V^{(N)}:=0$. This means that the picture $N$ is the minimum positive picture at which the operators $V \subset H_N$ can exist.

2) The negative ghost number $-N$ cohomology $H_{-N}$ consists of the physical vertex operators that exist at negative superconformal pictures $n \leq -N (N>0)$ and that are annihilated by the direct picture changing at maximum negative picture $-N$: $:\Gamma V^{(-N)}:=0$.

3) The operators existing at all pictures, including picture zero, at which they decouple from superconformal ghosts, are by definition the elements of the zero ghost cohomology $H_0$. The standard string perturbation theory thus involves the elements of $H_0$, such as a photon. The picture −3 and picture +1 five-forms considered above are the elements of $H_{-3}$ and $H_1$ respectively.

4) Generically, there is an isomorphism between the positive and the negative ghost cohomologies: $H_{-N-2} \sim H_N; N \geq 1$, as the conformal dimensions of the operators $: e^{-(N+2)\phi}(z)$ and $e^{N\phi}$ are equal and given by $-\frac{N^2}{2} - N$. That is, to any element of $H_{-N-2}$ there corresponds an element from $H_N$, obtained by replacing $e^{-(N+2)\phi} \to e^{N\phi}$ and adding the appropriate $b-c$ ghost terms in order to restore the BRST-invariance, using

the L-operator. For this reason, we shall refer to the cohomologies $H_{-N-2}$ and $H_N$ as dual. With some effort, one can also work out the precise isomorphism relation between $H_N$ and $H_{-N-2}$ That is, let the integrated vertex operator $h_{-N-2} \equiv \oint \frac{dz}{2i\pi} W_{-N-2}(z) \subset H_{-N-2}$ be the element of $H_{-N-2}$, with the integrand $W_{-N-2}(z)$ being some dimension 1 operator (for simplicity, we take it at tis maximal negative picture $-N-2$). The unintegrated version of this operator is given by $cW_{-N-2}$ Then the corresponding operator, the element of $H_N$: $h_N \equiv \oint \frac{dz}{2i\pi} W_N(z) \subset H_N$ can be constructed as

$$h_N =: (\Gamma^{2N+2} Z)^N cW_{-N-2} : \tag{3.14}$$

Since $Z$ and $\Gamma$ are BRST-invariant, the $h_N$-operator is also invariant by construction. In practice, the relation (3.14) is not convenient to work with because of the extreme complexity of manifest expressions involving powers of $Z$ and $\Gamma$-operators. One particularly efficient way of constructing the isomorphism involves the $K$-homotopy transform, that will be described in the next section. Our next step will be to take a closer look to the matter-ghost mixing space-time transformations of the type (3.3), (3.12) and to study the symmetry algebras they generate. This will particularly establish the relations between the ghost cohomologies, defined in this section, and the unbroken higher-spin symmetries in superstring theory.

## 3.2 AdS Isometries in Superstring Theory

Our purpose now is to point out the meaning of the space-time symmetries, described above. As we will show in this section, they are related to the hidden *AdS* isometries present in the RNS superstring theory (although this theory is originally defined in flat background). The $AdS_d$ isometry algebra is defined by the commutation relations:

$$\begin{aligned}
&[L^m, L^n] = -\frac{1}{\rho^2} L^{mn} \\
&[L^m, L^{np}] = -\eta^{mn} L^p + \eta^{mp} L^n \\
&[L^{mn}, L^{pq}] = -\eta^{np} L^{mq} - \eta^{mq} L^{np} + \eta^{mp} L^{nq} + \eta^{nq} L^{mp}
\end{aligned} \tag{3.15}$$

where $\rho$ is the *AdS* radius, related to *AdS* cosmological constant according to $\Lambda = -\rho^{-2}$ and the indices label the tangent space of $AdS_d$. In other words, the *AdS* isometry algebra (3.15) differs from the Poincare algebra, realizing the flat space-time isometries, by the noncommutation of the vector generators $L^m$ which, for this reason, are called the transvections, to distinguish them from the usual flat space-time translations.

In what follows, we shall identify the string-theoretic symmetry generators, structurally similar to (3.10), (3.11), to those generating AdS isometries (3.15). As previously, these generators are the elements of $H_1 \sim H_{-3}$. In turn, the space-time symmetry generators from the higher cohomologies form the covering of (3.15), leading to the full higher-spin algebra in $AdS$, realized in the larger string theory.

As the vertex operators, describing emissions of particles (e.g. a photon) by a string, originate from global space-time symmetry generators, the generators realizing $AdS$ isometries and their infinite-dimensional coverings, may similarly be used to deduce vertex operators in superstring theory, describing higher-spin dynamics in $AdS$. Such is the strategy that we will now follow.

As noted in the previous section, the space-time symmetry enerator $T^{(+)}$, although it is an integral of a dimension 1 primary, is not BRST-invariant, unlike its "dual" $T^{(-)}$. Namely, the BRST commutator with $T^{(+)}$ gives:

$$
\begin{aligned}
&[Q_{brst}, T^{(+)}] \\
&= \oint \frac{dz}{2i\pi}\{e^{2\phi-\chi} \times [-\frac{1}{2}\psi_n \partial X^n(\psi_m \partial^2 X^m - 2\partial\psi_m \partial X^m) \\
&+\frac{1}{6}\partial^3\psi_m\psi^m + \frac{1}{2}\partial^2\psi_m\partial\psi^m + \frac{1}{2}\partial^2 X_m \partial^2 X^m \\
&+\frac{1}{2}\partial^3 X_m \partial X^m + B^{(1)}_{\phi-\chi}(\frac{1}{2}\partial^2 X_m \partial X^m \\
&+\frac{1}{2}\partial^2\psi_m\psi^m + B^{(2)}_{\phi-\chi}\partial X_m \partial X^m)] \\
&+B^{(1)}_{2\phi-2\chi-\sigma}e^{3\phi-2\chi-\sigma}(\psi_m\partial^2 X^m - 2\partial\psi_m\partial X^m)\}
\end{aligned} \tag{3.16}
$$

To make this operator BRST-invariant (and thus an element of $H_1$) we will use a $K$-homotopy transformation, defined as follows.

Let

$$
T = \oint \frac{dz}{2i\pi} V(z) \tag{3.17}
$$

be an integral of a primary field of dimension 1, that does not commute with the supercurrent terms of the BRST operator, i.e. it is not BRST invariant, satisfying

$$
[Q_{brst}, V(z)] = \partial U(z) + W(z) \tag{3.18}
$$

and therefore

$$
[Q_{brst}, T] = \oint \frac{dz}{2i\pi} W(z) \tag{3.19}
$$

Introduce the homotopy operator

$$K(z) = -4ce^{2\chi-2\phi}(z) \equiv \xi\Gamma^{-1}(z) \tag{3.20}$$

satisfying

$$\{Q_{brst}, K(z)\} = 1 \tag{3.21}$$

In general, the homotopy operator has a non-singular OPE with $W$. Suppose this OPE is given by

$$K(z_1)W(z_2) \sim (z_1 - z_2)^N Y(z_2) + O((z_1 - z_2)^{N+1}) \tag{3.22}$$

where $N \geq 0$ and $Y$ is some operator of dimension $N+1$.

Then the complete BRST-invariant symmetry generator $L$ can be obtained from the incomplete non-invariant symmetry generator $T$ by the following homotopy transformation:

$$\begin{aligned} T \to L(w) = K\circ T = T + \frac{(-1)^N}{N!}\oint \frac{dz}{2i\pi}(z-w)^N :K\partial^N W:(z) \\ + \frac{1}{N!}\oint \frac{dz}{2i\pi}\partial_z^{N+1}[(z-w)^N K(z)]K\{Q_{brst}, U\} \end{aligned} \tag{3.23}$$

where $w$ is some arbitrary point on the worldsheet and we shall use the symbol $K\circ$ for the homotopy transformations to obtain BRST-invariant operators. It is straightforward to check the invariance of $K \circ T$ by using some partial integration as well as the obvious identity

$$\{Q_{brst}, W(z)\} = -\partial(\{Q_{brst}, U(z)\}) \tag{3.24}$$

that follows directly from (3.23). The homotopy transformed BRST-invariant operators are then typically of the form

$$K \circ T(w) = \oint \frac{dz}{2i\pi}(z-w)^N \tilde{V}_{N+1}(z) \tag{3.25}$$

with the conformal dimension $N+1$ operator $\tilde{V}_{N+1}(z)$ in the integrand satisfying

$$[Q_{brst}, \tilde{V}_{N+1}(z)] = \partial^{N+1}\tilde{U}_0(z) \tag{3.26}$$

where $\tilde{U}_0$ is some operator of conformal dimension zero. In particular, if $T$ is a space-time symmetry generator of the form

$$T = \oint \frac{dz}{2i\pi} e^{n\phi} F_{n^2+n+1} \tag{3.27}$$

where F is a matter primary of dimension $n^2+n+1$ and is annihilated by $\Gamma^{-1}$ at picture $n$, then the $K$-homotopy transformation maps $T$ into BRST-invariant element $K \circ T$ of $H_n$. It is straightforward to check that $K \circ T$ of $H_n$ is isomorphic to the space-time symmetry symmetry generator $\oint \frac{dz}{2i\pi} e^{-(n+2)\phi} F_{n^2+n+1}$ which is annihilated by $\Gamma$ and is an element of $H_{-n-2}$, with the explicit isomorphism relation given by (3.14).

The choice of the $w$-point in the transformation (3.22) is arbitrary since all the $w$-derivatives of $K \circ T$ are BRST-exact. Indeed, using $T(z) = \{Q, b\}$ and the invariance of $K \circ T$, it is straightforward to check that

$$\partial_w(K \circ T(w)) = \{Q, b_0 \partial_w K \circ T(w)\} \tag{3.28}$$

and similarly for the higher $w$-derivarives. Since the $BRST$-exact terms in the small Hilbert space do not contribute to the correlators, this ensures that all the correlation functions involving $K \circ T$ are independent on the choice of $w$. For our purposes, it will be also convenient to generalize the definition (3.22) as follows. Namely, we shall refer to operator $L = K_\Upsilon \circ T$ as a *partial* homotopy transform of $T$ based on $\Upsilon$, if the operator $T = \oint V$ satisfies $[Q, V] = \partial(cU) + W$, $\Upsilon$ is some dimension 1 operator, the OPE of $K$ and $\Upsilon$ is non-singular with the leading order $N > 0$ and $L$ is related to $K$ according to the transformation (3.22) with $W$ replaced by $\Upsilon$, i.e.

$$\begin{aligned} L(w) = K_\Upsilon \circ T = T + \frac{(-1)^N}{N!} \oint \frac{dz}{2i\pi}(z-w)^N : K\partial^N \Upsilon : (z) \\ + \frac{1}{N!} \oint \frac{dz}{2i\pi} \partial_z^{N+1}[(z-w)^N K(z)] K\{Q_{brst}, U\} \end{aligned} \tag{3.29}$$

Particularly, if $[Q, T] = \oint \Upsilon$, the partial homotopy transform obviously coincides with the usual homotopy transform (3.22). Having introduced all these definitions and prescriptions, we are now ready to construct the symmetry generators, realizing the AdS isometries in superstring theory, as well as the envelopings of these isometries, describing the higher-spin modes propagating in *AdS*. All these isometries and their envelopings will accordingly classified in terms of ghost cohomologies $H_n \sim H_{-n-2}$ of NSR superstring theory, *initially* defined in flat background. Let us start with the space-time symmetry generators of $H_1 \sim H_{-3}$. As we already pointed out, the scalar generators (3.10), (3.11) simply commutes with the standard Poincare generator and, in case of uncompactified critical string theory its geometrical meaning seems somewhat unclear. Let us, however, consider a noncritical theory in $D$ space-time dimensions, with $\varphi, \lambda$ parametrizing the super Liouville direction (alternatively, in critical superstring theory, one can consider one of the space-time dimensions compactified). It is convenient to cast the BRST charge according to

$$\begin{aligned}
Q &= Q_1 + Q_2 + Q_3 \\
Q_1 &= \oint \frac{dz}{2i\pi}(cT - bc\partial c) \\
Q_2 &= -\frac{1}{2}\oint \frac{dz}{2i\pi}(\gamma\psi_m\partial X^m - q\partial\lambda) \\
Q_3 &= -\frac{1}{4}\oint \frac{dz}{2i\pi}b\gamma^2
\end{aligned} \tag{3.30}$$

Consider the full (matter+ghost+Liouville) action (3.2). Apart from the space-time symetry transformations (3.3), mixing the matter and the ghosts, it is also invariant, in the limit of vanishing Liouville $2d$ cosmological constant, under the space-time transformations, which structure is similar to (3.12), but which mixes the matter and Liouville (alternatively, uncompactified and compactified) space-time directions:

$$\begin{aligned}
\delta X^m &= \epsilon^m\{\partial(e^\phi\lambda) + 2e^\phi\partial\lambda\} \\
\delta\lambda &= -\epsilon^m\{e^\phi\partial^2 X_m + 2\partial(e^\phi\partial X_m) \\
\delta\gamma &= \epsilon^m e^{2\phi-\chi}(\lambda\partial^2 X_m - 2\partial\lambda\partial X_m) \\
\delta\beta &= \delta b = \delta c = 0
\end{aligned} \tag{3.31}$$

The generators of these transformations are the $D$-vector components and the elements of $H_1 \sim H_{-3}$, given by:

$$T_m = \frac{1}{\rho}\oint \frac{dz}{2i\pi}e^\phi(\lambda\partial^2 X_m - 2\partial\lambda\partial X_m)$$

where $\rho$ is some constant (which we shall relate to AdS radius and cosmological constant, while relating $T_m$ to generator of transvections). This generator is not BRST-invariant and therefore the symmetry transformations generated by (3.30) are incomplete (similarly, the rotation generator $T_{mn} = \oint \frac{dz}{2i\pi}\psi_m\psi_n$ is not BRST invariant and therefore only induces rotations for the $\psi$-fields but not for bosons). To make both $T_m$ and $T_{mn}$ complete one has to restore their BRST invariance by adding ghost dependent correction terms. These terms can be obtained by the $K$-homotopy transformation, described above.

The homotopy transformed full BRST-invariant rotation generator is then given by

$$
\begin{aligned}
L_{mn} &= K \circ T_{mn} \\
&= \oint \frac{dz}{2i\pi}[\psi_m\psi_n + 2ce^{\chi-\phi}\psi_{[m}\partial X_{n]} - 4\partial cce^{2\phi-2\chi}] \\
&= -4\{Q, \xi \Gamma^{-1}\psi_m\psi_n\}
\end{aligned}
\tag{3.32}
$$

Note that the generator (3.31) can be written as a BRST commutator in the *large* Hilbert space. It is straightforward to check that the generator (3.31) induces (up to the terms, BRST exact in *small* Hilbert space) Lorenz rotations for all the matter fields (both $X$ and $\psi$). Similarly, the homotopy transformation of the vector generator gives full BRST-invariant symmetry generator given by

$$
\begin{aligned}
L^m(w) &= K \circ T^m \\
&= \oint \frac{dz}{2i\pi}(z-w)^2\{\frac{1}{2}B^{(2)}_{2\phi-2\chi-\sigma}e^{\phi}F^m_{\frac{5}{2}} - 12\partial cce^{2\chi-\phi}F^m_{\frac{5}{2}} \\
&+ce^{\chi}[-\frac{2}{3}\partial^3\psi^m\lambda + \frac{4}{3}\partial^3\varphi\partial X^m + 2\partial^2\psi^m\partial\lambda \\
&+B^{(1)}_{\phi-\chi}(-2\partial\varphi\partial^2X^m + 4\partial^2\varphi\partial X^m - 2\partial^2\psi^m\lambda + 4\partial\psi^m\partial\lambda) \\
&+B^{(2)}_{\phi-\chi}(2\partial\varphi\partial X^m + 2\psi^m\partial\lambda - 2\partial\psi^m\lambda - q\partial^2X^m) \\
&+B^{(3)}_{\phi-\chi}(-\frac{2}{3}\psi^m\lambda + \frac{4q}{3}\partial X^m)]\} \\
&= -4\{Q, \oint\frac{dz}{2i\pi}(z-w)^2ce^{2\chi-\phi}F^m_{\frac{5}{2}}(z)\}
\end{aligned}
\tag{3.33}
$$

so the full vector symmetry generator is again the BRST commutator in the large Hilbert space. Here $F^m_{\frac{5}{5}} = \lambda\partial^2X_m - 2\partial\lambda\partial X_m$ The BRST-invariant symmetry generator can also be be constructed at dual $-3$ picture (as well as the pictures below; but not above minimal negative picture $-3$ at which it is annihilated by the picture changing). At picture $-3$ the related symmetry generator is given by

$$
T^{m(-3)} = \oint \frac{dz}{2i\pi}e^{-3\phi}F^m_{\frac{5}{2}}
\tag{3.34}
$$

Now we can show that $L^m$ and $L^{mn}$-generators form AdS isometry algebra. To demonstrate this, we start with the OPE of the the primary fields $F^m_{\frac{5}{2}}(z)$

(related to the matter ingredient of $L^m$). Straightforward calculation gives

$$\begin{aligned}
&F^m_{\frac{5}{2}}(z)F^n_{\frac{5}{2}}(w) = -\frac{6\eta^{mn}}{(z-w)^5} + \frac{14\partial\lambda\lambda\eta^{mn}(w) + 8\partial X^m\partial X^n(w)}{(z-w)^3}\\
&+\frac{10\partial^2 X^m\partial X^n(w) - 2\partial X^m\partial^2 X^n(w) + 7\eta^{mn}\partial^2\lambda\lambda(w)}{(z-w)^2}\\
&+\frac{6\partial^3 X^m\partial X^n(w) - 3\partial^2 X^m\partial^2 X^n(w)}{z-w}\\
&+\frac{3\eta^{mn}\partial^3\lambda\lambda(w) + 2\eta^{mn}\partial^2\lambda\partial\lambda(w)}{z-w}\\
&+(z-w)^0[\frac{7}{3}\partial^4 X^m\partial X^n(w) - 2\partial^3 X^m\partial^2 X^n(w)\\
&+\frac{11}{12}\eta^{mn}\partial^4\lambda\lambda(w) + \frac{4}{3}\eta^{mn}\partial^3\lambda\partial\lambda(w)] + :F^m_{\frac{5}{2}}F^m_{\frac{5}{2}}:(w)\\
&+(z-w)[\frac{13}{60}\eta^{mn}\partial^5\lambda\lambda(w) - \frac{1}{2}\eta^{mn}\partial^4\lambda\partial\lambda(w) + \frac{2}{3}\partial^5 X^m\partial X^n(w)\\
&-\frac{5}{6}2\partial^4 X^m\partial^2 X^n(w) + :\partial F^m_{\frac{5}{2}}F^m_{\frac{5}{2}}:(w)]\\
&+(z-w)^2[\frac{1}{24}\eta^{mn}\partial^6\lambda\lambda(w) - \frac{2}{15}\eta^{mn}\partial^5\lambda\partial\lambda(w)\\
&+\frac{3}{20}\partial^6 X^m\partial X^n(w) - \frac{1}{4}\partial^5 X^m\partial^2 X^n(w) + \frac{1}{2}:\partial^2 F^m_{\frac{5}{2}}F^m_{\frac{5}{2}}:(w) + \ldots
\end{aligned}\tag{3.35}$$

Using this OPE it is straightforward to compute the commutator $[L^m, L^n]$. It is convenient to choose one of the vector at picture $+1$ representation (3.32) and another at negative picture $-3$ representation. Because of the isomorphism between positive and negative picture representations, ensured by the appropriate $Z, \Gamma$ transformations, the final result will be picture-independent. Then, using (3.34) and the BRST invariance of $L^m$, we get

$$\begin{aligned}
&[L^m, L^n]\\
&= \frac{1}{\rho^2}\{Q, [\oint\frac{dz_1}{2i\pi}(z_1-w)^2 ce^{2\phi-\chi}F^m_{\frac{5}{2}}(z_1), \oint\frac{dz_2}{2i\pi}e^{-3\phi}F^m_{\frac{5}{2}}(z_2)]\}\\
&= \{Q, U(z_2)\}
\end{aligned}\tag{3.36}$$

where

$$
\begin{aligned}
U(z_1) \equiv & \oint \frac{dz_2}{2i\pi} U_1(z_2) + \oint \frac{dz_2}{2i\pi}(z_1 - z_2)U_2(z_2) \\
& + \oint \frac{dz_2}{2i\pi}(z_1 - z_2)^2 U_3(z_2) \\
& = \oint \frac{dz_2}{2i\pi} ce^{2\chi-4\phi}[\frac{7}{3}\partial^4 X^{[m}\partial X^{n]} - 2\partial^3 X^{[m}\partial^2 X^{n]} \\
& +6P^{(1)}_{2\chi-\phi+\sigma}\partial^3 X^{[m}\partial X^{n]} \\
& +4P^{(2)}_{2\chi-\phi+\sigma}\partial^3 X^{[m}\partial X^{n]} + :F^m_{\frac{5}{2}}F^m_{\frac{5}{2}}:] \\
& + \oint \frac{dz_2}{2i\pi}(z_1 - z_2)ce^{2\chi-4\phi}[\frac{4}{3}B^{(3)}_{2\chi-\phi+\sigma}\partial^2 X^{[m}\partial X^{n]} \\
& +3B^{(2)}_{2\chi-\phi+\sigma}\partial^3 X^{[m}\partial X^{n]} \\
& +B^{(1)}_{2\chi-\phi+\sigma}(\frac{7}{3}\partial^4 X^{[m}\partial X^{n]} - 2\partial^3 X^{[m}\partial^2 X^{n]}) \\
& +\frac{2}{3}\partial^5 X^{[m}\partial X^{n]} - \frac{5}{6}\partial^4 X^{[m}\partial^2 X^{n]}] \\
& + \oint \frac{dz_2}{2i\pi}(z_1 - z_2)^2 ce^{2\chi-4\phi}[\frac{1}{3}B^{(4)}_{2\chi-\phi+\sigma}\partial^2 X^{[m}\partial X^{n]} \\
& +B^{(3)}_{2\chi-\phi+\sigma}\partial^3 X^{[m}\partial X^{n]} + B^{(2)}_{2\chi-\phi+\sigma}(\frac{7}{6}\partial^4 X^{[m}\partial X^{n]} - \partial^3 X^{[m}\partial^2 X^{n]}) \\
& +B^{(1)}_{2\chi-\phi+\sigma}(\frac{2}{3}\partial^5 X^{[m}\partial X^{n]} - \frac{5}{6}\partial^4 X^{[m}\partial^2 X^{n]}) \\
& +\frac{3}{20}\partial^6 X^{[m}\partial^2 X^{n]} - \frac{1}{4}\partial^5 X^{[m}\partial^2 X^{n]}]
\end{aligned}
\tag{3.37}
$$

where, for convenience, we split the overall integral into 3 parts, with the integrands proportional to $U_1(z_2)$, $(z_1-z_2)U_2(z_2)$ and $(z_1-z_2)^2U(z_2)$ accordingly. To relate the right hand side of the commutator (3.36) to the rotation generator (3.31) one has to perform double picture changing transform of $\{Q, U(z_1)\}$ in order to bring it to picture zero. We shall demonstrate the procedure explicitly for the $U_1$ integral, with the other two integrals treated similarly. For that, we first of all need a manifest expression for the commutator of the BRST charge with the $U_1(z_1)$ operator (3.36). Straightforward calculation gives:

$$\{Q,U_1(z)\} = -2\partial cce^{2\chi-4\phi}(B^{(2)}_{2\chi-\phi+2\sigma} + B^{(2)}_{2\chi-\phi+\sigma})\partial^2 X^{[m}\partial X^{n]}$$
$$+9\partial^2 cce^{2\chi-4\phi}B^{(1)}_{2\chi-\phi+\sigma}\partial^2 X^{[m}\partial X^{n]}$$
$$-\frac{3}{2}\partial^2 cce^{2\chi-4\phi}\partial^3 X^{[m}\partial X^{n]}$$
$$+\partial cce^{2\chi-4\phi}(\frac{7}{3}\partial^4 X^{[m}\partial X^{n]} - 2\partial^3 X^{[m}\partial X^{n]})$$
$$+\frac{34}{3}\partial^3 cce^{2\chi-4\phi}\partial^2 X^{[m}\partial X^{n]} + 12\partial^2 cce^{2\chi-4\phi}\partial^3 X^{[m}\partial X^{n]}$$
$$+4ce^{\chi-3\phi}[\partial^2\psi^{[m}\partial X^{n]} + \partial\psi^{[m}\partial^2 X^{n]}$$
$$-B^{(1)}_{\phi-\chi}(\psi^{[m}\partial^2 X^{n]} - 2\partial\psi^{[m}\partial X^{n]})$$
$$+\psi^{[m}\partial X^{n]}(B^{(2)}_{\phi-\chi} + B^{(2)}_{2\chi-\phi+\sigma})] \tag{3.38}$$

The next step is to perform the normal ordering of the integrand of this expression with $\xi = e^{\chi}$ around the midpoint. We get

$$:\xi\{Q,U_1(z)\}:= -8\partial cce^{3\chi-4\phi}\partial^2 X^{[m}\partial X^{n]} + 4ce^{2\chi-3\phi}[\psi^{[m}\partial^2 X^{n]}$$
$$+4\partial\psi^{[m}\partial X^{n]} - 2\psi^{[m}\partial X^{n]}P^{(1)}_{3\phi-\chi-2\sigma}] \tag{3.39}$$

The next step is to perform the commutation of this expression with $Q$ which, by definition, gives us $\{Q,U(z_1)\}$ at picture $-1$. Straightforward calculation gives:

$$\{Q,\xi\{Q,U_1(z)\}\} = ce^{\chi-2\phi}(4B^{(1)}_{\phi+\chi-2\sigma}\psi^m\psi^n + 2\psi^{[m}\partial\psi^{n]}) \tag{3.40}$$

Next, the normal ordering of this expression with $\xi$ around the midpoint gives

$$:\xi\{Q,\xi\{Q,U_1(z)\}\}:= 4ce^{2\chi-2\phi}\psi^m\psi^n \tag{3.41}$$

Finally, the commutator of this expression with Q by definition gives us $U_1$ at picture zero:

$$U_1^{(0)}(z) \equiv \{Q,\xi\{Q,\xi\{Q,U_1(z)\}\}\} = 4\{Q,e^{2\chi-2\phi}\psi^m\psi^n\} \tag{3.42}$$

which, according to (3.31) is nothing but the integrand of the full rotation generator with the inverse sign. The picture transform of $U_2$ and $U_3$ in the remaining terms of (3.36) is performed similarly. Applying picture changing transformation twice and integrating out total derivatives we find the contributions from the second and the third integrals cancel each other and the commutator remains unchanged. This concludes the proof that the commutator of two operators $[L^m, L^n] = -\frac{1}{\rho^2}L^{mn}$ reproduces the commutation of two transvections in the *AdS* isometry algebra.

The remaining commutators realizing the AdS isometries, are computed similarly. Note the highly nontrivial appearance of the minus sign on the right hand side of the commutator as a result of the ghost structure of the operators, that indicates that the effective cosmological constant in the symmetry algebra is negative, so the effective geometry generated is of the *AdS* type.

The combination of operators $(L^m, L^{mn})$, that we considered so far, is not the only possible realization of the AdS symmetry algebra in RNS theory. In particular, it is easy to check that so(d-1, 2) isometry of the AdS space is realized by $S^m, L^{mn}$ where

$$L^{mn} = K \circ T^{mn} \equiv K \circ \oint \frac{dz}{2i\pi}\psi^m\psi^n \tag{3.43}$$

is the same full rotation operator (where the $K\circ$ represents the homotopy transformation to ensure the BRST-invariance) while $S^m$ is the homotopy transformation of the operator $\oint \frac{dz}{2i\pi}\lambda\psi^m$, representing the rotation in the Liouville-matter plane:

$$\begin{aligned} S^m &= K \circ \rho^{-1} \oint \frac{dz}{2i\pi}\lambda\psi^m \\ &= \rho^{-1} \oint \frac{dz}{2i\pi}[\lambda\psi^m + 2ce^{\chi-\phi}(\partial\varphi\psi^m - \partial X^m\lambda - qP^{(1)}_{\phi-\chi}\psi^m) \\ &-4\partial cce^{2\chi-2\phi}\lambda\psi^m] \\ &= -4\{Q, \rho^{-1} \oint \frac{dz}{2i\pi}ce^{2\chi-2\phi}\lambda\psi^m\} \end{aligned} \tag{3.44}$$

Again, with the procedure identical to the one explained above, it is straightforward to show that $S^m$ satisfy the commutation relation for transvections: $[S^m, S^n] = -\frac{L^{mn}}{\rho^2}$ and the rest of $so(d-1, 2)$ relations with $L^{mn}$. In order to construct vertex operators for spin connection in *AdS* space, we will actually need the realization using the linear combination of the transvections (3.33), (3.44), given by

$$P^m = \frac{1}{\sqrt{2}}(L^m + S^m) \tag{3.45}$$

One can show that these generators realize the transvections on $AdS$ space provided that the two-form $L^{mn}$ is shifted according to

$$\begin{aligned}
&L^{mn} \to P^{mn}(w) \\
&= L^{mn} + K \circ \oint \frac{dz}{2i\pi} e^{-3\phi}(\psi^{[m}\partial^2 X^{n]} - 2\partial\psi^{[m}\partial X^{n]}) \\
&+qK \circ \oint \frac{dz}{2i\pi} ce^{\chi-\phi}\lambda\psi^m\psi^n
\end{aligned} \tag{3.46}$$

where $K$ is again the homotopy transformation. The new term, proportional to the background charge, appears as a result of the Liouville terms of $Q$ entering the game of picture changing. The AdS isometry algebra is then realized by the combination of $P^m$ and $P^{mn}$. In the next section, we will use these $AdS$ isometry generators as building blocks to construct massless vertex operators for $AdS$ frame-line higher spin fields in closed and open string theory. The closed string sector particularly turns out to contain a graviton mode propagating in AdS geometry. In the open string sector, we will relate the vertex operators of $H_n \sim H_{-n-2}$ to the realizations of infinite-dimensional $AdS$ higher-spin symmetries.

## 3.3 Vertex Operators for Frame-Like Higher-Spin Fields and AdS Higher-Spin Algebra

The Poincare space-time symmetry algebra is induced by generators (elements of $H_0$), emerging in the zero-momentum limit of massless open string vertex operator (a photon). In closed string theory, the corresponding massless vertex operators (a graviton, an axion and a dilaton) are the bilinears in these symmetry generators (the elements of $H_0 \otimes H_0$). Following this pattern, we shall now discuss the new classes of the vertex operators related to the space-time symmetries (3.15) realizing the $AdS$ isometry algebra (elements of $H_1 \sim H_{-3}$) and its higher-spin coverings (involving the elements of $H_1 \otimes H_{-3}$ and higher cohomologies). In conventional string theory, the physical vertex operators are usually regarded as sources of space-time fields, such as a graviton (space-time metric). The low-energy limit of string theory (obtained from vanishing $\beta$-function conditions) then reproduces the Einstein-Hilbert theory of gravity, with the Einstein equations involving the scalar curvature and Ricci tensor, constructed out of derivatives of the metric. Similarly, the symmetric fields in the Fronsdal's actions (1.1), (1.5) are the higher-spin analogues of the metric tensor, with the Lagrangians containing linearized higher-spin curvatures. In this sense, both the Einstein-Hilbert formulation of gravity and the Fronsdal's formulation of its (linearized) higher-spin extensions are known as *metric-like* formulations of the gravity and its higher-spin generalizations. In the alternative formulation, known as the frame, or Cartan-Weyl formulation of gravity (or Mac Dowell-Mansouri-Stelle-West (MMSW) formulation the case of nonzero cosmological constant) one starts with the vielbein field $e_m^a$ (where $m$ labels the curved manifold and $a$ labels the tangent space) and the spin connection $\omega_m^{ab}$, which is antisymmetric in the tangent space. $\omega$ and $e$ are not independent fields are not independent. For the sake of certainty, below we consider the case of nonzero (negative) cosmological constant. Introduce the Cartan-Weyl one-form

$$\Omega^{(1)} \equiv \Omega_m dx^m = e_m^a T_a dx^m + \omega_m^{ab} T_{ab} dx^m \tag{3.47}$$

where $T_a$ and $T^{ab}$ are the $AdS$ isometry generators. Differentiating $\Omega^{(1)}$ using the commutation relations of the isometry algebra (3.15) , we obtain the two-form

$$d\Omega^{(1)} = (K_{mn}^a T_a + R_{mn}^{ab} T_{ab}) dx^m \wedge dx^n \tag{3.48}$$

which components define the torsion and the Lorentz curvature according to

$$\begin{aligned} R^{ab} &= d\omega^{ab} + (\omega \wedge \omega)^{ab} + \Lambda (e \wedge e)^{ab} \\ K^a &= de^a + (\omega \wedge e)^a \end{aligned} \tag{3.49}$$

Imposing the zero torsion condition $K = 0$ relates $\omega$ and $e$. This condition can be resolved expressing the spin connection $\omega$ in terms of the vielbein field: $\omega = \omega(e, \partial e)$. This implies that the only true dynamical field in the MMSW formalism is the vielbein, while the spin connection is the *extra field*, expressed in the derivatives of $e$. This construction is straightforward to generalize for the higher-spin case. For simplicity, below we shall restrict ourselves to the case of symmetric fields (the mixed-symmetry case can also be considered). To describe a spin $s$ symmetric field $H_{m_1 \dots m_s}$ in the frame-like formalism, introduce a set of two-row fields $\Omega_{s-1|t} \equiv \Omega_m^{a_1 \dots a_{s-1}|b_1 \dots b_t}; 0 \le t \le s-1$, symmetric in $a$ and $b$-indices.

The only dynamical field in this set is $\Omega_{s-1|0}$ ($t = 0$), expressed in the tangent space as $\Omega^{a|a_1 \dots a_{s-1}}(x) = e_m^a(x) \Omega^{m|a_1 \dots a_{s-1}}$. This field is now a sum of two irreducible fields, one being a rank $s$ symmetric field $\Omega^{(a|a_1 \dots a_{s-1})}$ and another a two-row field (with the $a$-index making the second row of unit length). As in the case of the MMSW gravity, the $t = 0$ field with two rows can be removed by partially fixing the gauge symmetry

$$\Omega_m^{a_1 \dots a_{s-1}} \to \Omega_m^{a_1 \dots a_{s-1}} + D_m \Lambda^{a_1 \dots a_{s-1}} \tag{3.50}$$

(wih the covariant derivative taken around the *AdS* vacuum). Then the metric higher-spin field $H$ is related to $\Omega$ according to

$$H_{m_1....m_s} = e^{a_1}_{m_1}....e^{a_s}_{m_s}\Omega^{a_1...a_s} \tag{3.51}$$

Just as the spin connection, the $t \neq 0$ frame-like fields are the extra fields, related to the dynamical field $\Omega^{s-1|0}$ through the generalized zero-curvature conditions:

$$\Omega^{s-1|t}(x) \sim \partial\Omega^{s-1|t-1}(x) \sim \partial^t\Omega^{s-1|0} \tag{3.52}$$

The explicit form of the zero-torsion constraints depends is determined by the higher-spin extensions of curvature and torsion 2-forms which, beyond the linearized limit, are not simple objects to work with. There is, however, a way to extract these constraints explicitly from string theory (see below).

With the higher-spin gauge fields in the frame-like formalism, the Cartan-Weyl 1-form is extended according to

$$\begin{aligned}\Omega^{(1)}_{H.S.} &= (e^a_m T_a + \omega^{ab}_m T_{ab}\\ &+\sum_{s=3}^{\infty}\sum_{t=0}^{s-1}\omega^{a_1...a_{s-1}|b_1...b_t}_m T_{a_1...a_{s-1}|b_1...b_t})dx^m\end{aligned} \tag{3.53}$$

where $T_{a_1...a_{s-1}|b_1...b_t} \equiv T_{s|t}$ realize the $AdS_d$ higher-spin algebra and form a particular covering of $SO(2, d-1)$. The commutators are structurally given by

$$[T_{s_1|t_1}, T_{s_2|t_2}] = \sum_{s_3=|s_1-s_2|}^{s_1+s_2-1}\sum_{t_3=0}^{s_3-1} C^{s_3,t_3}_{s_1,t_1|s_2,t_2}T_{s_3,t_3} \tag{3.54}$$

where the structure constants depend, in general, on a choice of the basis and on concrete realization of the covering of the *AdS* isometry algebra. In particular, as it is clear from the underlying *AdS* isometry,there exists a choice of the basis such that the structure constants have the schematic form:

$$\begin{aligned}&C = K_{SO(2,d-1)}[\{\lambda\}(s_1,t_1); \{\lambda\}(s_2,t_2); \{\lambda\}(s_3,t_3));\\ &f(\{\rho\}(s_1), \{\rho\}(s_2), \{\rho\}(s_3)]\end{aligned} \tag{3.55}$$

where $K$ are $SO(2, d-1)$ Clebsch-Gordan coefficients (with $\{\lambda\}$ labelling the vectors in irreducible representations and $f$ being a certain function of Casimir eigenvalues (labeled by $\{\rho\}$). So formally, the nontrivial part of the higher-spin algebra is encrypted in the undetermined function $f(\rho)$, although the basis choice related to (3.55) isn't always convenient for practical purposes. Explicit examples of higher-spin algebra in both flat and *AdS* space-time geometries have been studied extensively and the higher-spin structure constants define certain classes of cubic higher-spin interactions and are holographically dual to the CFT structure constants in the dual theories on $AdS$ boundaries. As it is clear from (3.53), the higher spin algebra (an the interactions it describes) cannot be truncated at levels beyond spin 2: e.g. a collision of spin 3 particles would produce, in particular, a particle of spin 4; interaction of spins 3 and 4 produces spin 5, and so on. In other words, as soon as we go beyond the realm of lower-spin physics (with spin values $s \leq 2$), there is no way to limit ourselves to, say, spin values of 3 or 4 - we have to face the whole infinite tower of higher-spin modes. The higher-spin algebras are essentially infinite-dimensional, with $SO(2, D-2)$ being their maximal finite subalgebras. As we will see in the next chapter, this is precisely the situation reproduces in the larger string theory with the nonzero ghost cohomologies included; the latter actually hold the keys to the unbroken higher-spin symmetries.

# Chapter 4

# Interactions of Higher Spin Fields in Larger String Theory

## 4.1 Vertex Operators in the Frame-Like Formalism

As we can see from the previous discussion, the physical massless vertex operators in string theory are related to space-time isometry generators (e.g. a photon and a graviton constructed out of Poincare generators). In the standard approach, these operators (e.g. a photon) are the sourses of the metric-like fields (such as $g_{mn}$ in the target space). As we have seen in the previous section, however, the higher-spin extensions of the space-time symmetries (involving the $T_{s|t}$-generators are most naturally realized in the frame-like description. In particular, the $AdS$ isometry algebra (3.15) is realized in $RNS$ superstring theory through operators of nonzero ghost cohomology $H_1 \sim H_{-3}$. This algebra does not mix with Poincare isometries of $H_0$ , as the $H_1 \sim H_{-3}$ generators all commute with those of Poincare algebra, up to BRST-exact terms. The higher-spin extension of the $AdS$ isometries will then turn out to come from the higher ghost cohomologies $H_{s-2} \sim H_{-s}$, with $s$ corresponding to the spin value. The structure (3.54) of the higher-spin algebra is given by the OPE fusion rules for the cohomologies in this language, with the structure constants of the higher-spin algebra read off the three-point correlators of the higher-spin operators. Based on that, our strategy now will be to first construct the vertex operators in $H_1 \sim H_{-3}$ that will be the sourses of the frame fields (vielbein gauge fields) in space-time, with the low-energy limit of the corresponding sigma-model reproducing the MMSW gravity. In the closed string sector,these operators will particularly describe the gravitons propagating in the $AdS$ space-time. Then, considering the higher cohomologies, $H_{s-2} \sim H_{-s}$, we shall construct the vertex operators for the higher-spin gauge fields in the frame-like formalism. The BRST-invariance conditions for these operators will lead to the on-shell =equations of motion in $AdS$ space-time. The gauge symmetries will stem ffrom the BRST nontriviality.

Conditions on the operators, e.g. the gauge transformations for the frame-like higher-spin fields will be shifting the operators by the BRST-exact terms. Once the equations of motion/gauge symmetries of the higher-spin fields are fixed by the BRST properties of the vertex operators, the correlation functions of these operators, computed on the worldsheet, are gauge-invariant by constructions and can be used to establish the form of the higher-spin interactions in the $AdS$ space. So we start from the graviton's vertex operator in closed string theory, describing the graviton's dynamics in the $AdS$ space in the MMSW formalism. Based on the $AdS$ isometry generators (3.33), (3.45) the proposed operator is

$$G(p) = e_m^a(p)F_a\bar{L}^m + +\omega_m^{ab}(p)(F_b^m\bar{L}_a - \frac{1}{2}F_{ab}\bar{L}^m) + c.c. \tag{4.1}$$

where

$$\begin{aligned} F_m &= -2K_{U_1} \circ \int dz\lambda\psi_m e^{ipX}(z) \\ U_1 &= \lambda\psi_m e^{ipX} + \frac{i}{2}\gamma\lambda((\vec{p}\vec{\psi})\psi_m - p_m P^{(1)}_{\phi-\chi})e^{ipX} \end{aligned} \tag{4.2}$$

or manifestly

$$\begin{aligned} F_m &= -2\int dz\{\lambda\psi_m(1 - 4\partial cce^{2\chi-2\phi}) + \\ &2ce^{\chi-\phi}(\lambda\partial X_m - \partial\varphi\psi_m + q\psi_m P^{(1)}_{\phi-\chi} \\ &-\frac{i}{2}((\vec{p}\vec{\psi})\psi_m - p_m P^{(1)}_{\phi-\chi}))\}e^{ipX}(z) \end{aligned} \tag{4.3}$$

Next,

$$\begin{aligned} \bar{L}^a &= \int d\bar{z}e^{-3\bar{\phi}}\{\bar{\lambda}\bar{\partial}^2 X^a - 2\bar{\partial}\bar{\lambda}\bar{\partial}X^a \\ &+ip^a(\frac{1}{2}\bar{\partial}^2\bar{\lambda} + \frac{1}{q}\bar{\partial}\bar{\varphi}\bar{\partial}\bar{\lambda} - \frac{1}{2}\bar{\lambda}(\bar{\partial}\bar{\varphi})^2 \\ &+(1+3q^2)\bar{\lambda}(3\bar{\partial}\bar{\psi}_b\bar{\psi}^b - \frac{1}{2q}\bar{\partial}^2\bar{\varphi}))\}e^{ipX} \end{aligned} \tag{4.4}$$

(similarly for its holomorphic counterpart $L^a$) and

$$F_{ma} = F^{(1)}_{ma} + F^{(2)}_{ma} + F^{(3)}_{ma} \tag{4.5}$$

where

$$\begin{aligned} F^{(1)}_{ma} &= -4qK_{U_2} \circ \int dzce^{\chi-\phi}\lambda\psi_m\psi_a \\ U_2 &= [Q - Q_3, ce^{\chi-\phi}\lambda\psi_m\psi_a e^{ipX}] \\ &-\frac{i}{2}c\lambda((\vec{p}\vec{\psi})\psi_a\psi_m - p_m\psi_a P^{(1)}_{\phi-\chi})e^{ipX}(z) \end{aligned} \tag{4.6}$$

$$F_{ma}^{(2)} = K \circ \int dz\psi_m\psi_a e^{ipX} = -4\{Q, \int dzce^{2\chi-2\phi}e^{ipX}\psi_m\psi_a(z)\} \tag{4.7}$$

and

$$F_{ma} = F_{ma}^{(1)} + F_{ma}^{(2)} + F_{ma}^{(3)} \tag{4.8}$$

where

$$\begin{aligned} &F_{ma}^{(1)} = -4qK_{U_2} \circ \int dzce^{\chi-\phi}\lambda\psi_m\psi_a \\ &U_2 = [Q - Q_3, ce^{\chi-\phi}\lambda\psi_m\psi_a e^{ipX}] \\ &-\frac{i}{2}c\lambda((\vec{p}\vec{\psi})\psi_a\psi_m - p_m\psi_a P_{\phi-\chi}^{(1)})e^{ipX}(z) \end{aligned} \tag{4.9}$$

$$F_{ma}^{(2)} = K \circ \int dz\psi_m\psi_a e^{ipX} = -4\{Q, \int dzce^{2\chi-2\phi}e^{ipX}\psi_m\psi_a(z)\} \tag{4.10}$$

and

$$F_{ma}^{(3)} = \int dze^{-3\phi}(\psi_{[m}\partial^2 X_{a]} - 2\partial\psi_{[m}\partial X_{a]})e^{ipX}(z) \tag{4.11}$$

In the limit of zero momentum the holomorphic and the antiholomorphic components of the operator (4.1) correspond to AdS isometry generators (3.15), (3.33), (3.45) in different realizations, described above. More precisely, while the antiholomorphic part of (4.1) is based on the L- operators (3.33), related to the $L$-realization of the symmetry algebra, the holomorphic part of (4.1) involves the F-operators (such as $F^a$ and $F^{ab}$) which, although different from the operators (3.45), (3.47) of the $P$-representation, become related to those after one imposes the on-shell constraints on the space-time fields.

We start with analyzing the BRST invariance constraints on the operator (4.1). The BRST commutators are given by:

$$\begin{aligned} &[\bar{Q}, G(p)] = 0 \\ &[\bar{Q}, G(p)] = ie_m^b(p)\bar{L}_b \int dz\gamma\lambda((\vec{p}\vec{\psi})\psi_m - P_{\phi-\chi}^{(1)}p_m)e^{ipX}(z) \\ &+\omega_m^{ab}\bar{L}_b \int dz\{\gamma\lambda\psi_a\psi_m + 2ic\lambda(\vec{p}\vec{\psi})\psi_a\psi_m - P_{\phi-\chi}^{(1)}p_m\psi_a\}e^{ipX}(z) \end{aligned} \tag{4.12}$$

The BRST invariance therefore imposes the following constraints on vielbein and connection fields:

$$
\begin{aligned}
&p^{[n}e^{b}_{m]}(p) - \omega^{b[n}_{m]}(p) = 0\\
&p_{[n}\omega^{ab}_{m]}(p) = 0\\
&p^{m}e^{b}_{m}(p) = 0\\
&p^{m}\omega^{ab}_{m}(p) = 0
\end{aligned}
\tag{4.13}
$$

The first two constraints represent the linearized equations $R^{AB} = 0$ (the first one being the zero torsion constraint $T^a = R^{a\hat{d}} = 0$ whle the second reproducing vanishing Lorenz curvature $R^{ab} = 0$). The last two constraints represent the gauge fixing conditions related to the diffeomorphism symmetries. The fact that the BRST invariance leads to space-time equations in a certain gauge is not surprising if we recall that similar constraints on a standard vertex operator of a photon also lead to Maxwell's equations in the Lorenz gauge. Provided that the constraints (4.13) are satisfied, the vertex operator $G(p)$ can be written as a BRST commutator in the large Hilbert space plus terms that are manifestly in the small Hilbert space, according to

$$
\begin{aligned}
G(p) &= \{Q, W(p)\}\\
&+\frac{1}{q}\omega^{ab}_{m}\int dze^{-3\phi}(\psi^{[m}\partial^2 X_{a]} - 2\partial\psi_{[m}\partial^{X}_{a]})e^{ipX}(z)\bar{L}_b + c.c.\\
W(p) &= 8e^{a}_{m}(p)\bar{L}_a\int dzc\partial\xi\xi e^{-2\phi}\lambda\psi^{m}e^{ipX}\\
&+\omega^{ab}_{m}\bar{L}_b[-\frac{4}{q}\int dzc\partial\xi\xi e^{-2\phi}\psi_a\psi^{m}e^{ipX}\\
&+4\int dz(z-w)\partial cc\partial^2\xi\partial\xi\xi e^{-3\phi}\lambda\psi_a\psi^{m}e^{ipX}]
\end{aligned}
\tag{4.14}
$$

This particularly implies that, modulo gauge transformations, the vertex operator $G(p)$ is the element of the *small* Hilbert space. Let us now turn to the question of BRST nontriviality and related gauge symmetries (3.50). The linearized gauge symmetry transformations (3.50) are given by

$$
\begin{aligned}
\delta e^{a}_{m} &= \partial_m\rho^{a} + \rho^{a}_{m}\\
\delta\omega^{ab}_{m} &= \partial_m\rho^{ab} + \rho^{[a}\delta^{b]}_{m}
\end{aligned}
\tag{4.15}
$$

Where we write $\rho^{AB} = (\rho^{ab}, \rho^{a\hat{d}}) = (\rho^{ab}, \rho^{a})$. The variation of $G(p)$ under (4.15) in the momentum space is

$$
\delta G(p) = p^{m}F_{m}\bar{L}_a\rho^{a} + p^{m}F_{ma}\bar{L}_b\rho^{ab}
\tag{4.16}
$$

The two terms of the variation (4.16) are BRST exact in the *small* Hilbert space (and therefore are irrelevant in correlators) since

$$p^m F_m = \{Q, :\Gamma:(w)[Q, \xi A]\}$$
$$A = \int dz e^{\chi - 3\phi} \partial\chi((\vec{p}\partial\vec{X})\lambda - (\vec{p}\vec{\psi})\partial\varphi + (\vec{p}\vec{\psi})P^{(1)}_{\phi-(1+q)\chi})e^{ipX} \tag{4.17}$$

and

$$p^m F^{(1)}_{ma} = 4q[Q, \Gamma(w)\int dz c e^{-3\phi}\partial\xi\partial^2\xi\lambda\psi_a(\vec{p}\vec{\psi})e^{ipX}]$$
$$p^m F^{(2)}_{ma} = \{Q, :\Gamma:(w)\int dz\partial\xi e^{-3\phi}((\vec{p}\vec{\psi})\partial X_a - (\vec{p}\partial\vec{X})\psi_a)e^{ipX}\}$$
$$p^m F^{(3)}_{ma} = \{Q, [K \circ \int dz\lambda\psi_a e^{ipX}, B]\}$$
$$B = \int dz\partial\xi e^{-4\phi}[\lambda(\partial\vec{\psi}\partial^2\vec{X}) - 2\partial\lambda((\vec{\psi}\partial^2\vec{X}) - 2(\partial\vec{\psi}\partial\vec{X}))] \tag{4.18}$$

Therefore gauge transformations of $e$ and $\omega$ shift $G(p)$ by terms not contributing to correlators. This concludes the BRST analysis of the vertex operator for vielbein and connection fields in the frame-like description of MMSW gravity. Now we are prepared to investigate the conformal beta-function of $G(p)$ in the sigma-model, showing that it reproduces the equations of motion of MMSW gravity with negative cosmological constant in the low energy limit.

The leading order contribution to the beta-function of the $G(p)$ operator (giving the equations of motion for $e$ and $\omega$ in the low energy limit of string theory) is determined by the structure constants stemming from three-point correlators on the worldsheet. Computing these structure constants will be our goal in this section. Manifest expressions for the operators (4.1)-(4.11) look quite lengthy and complicated. The computations, however, can be simplified significantly due to important property of the homotopy transformation (3.23): That is, consider two operators $V_1(z)$ and $V_2(w)$ (of dimension 1) that are, in general, not BRST-invariant and are the elements of the small space (i.e. independent on zero mode of $\xi$). Suppose their operator products with the homotopy operator $K$ are nonsingular while their full OPE between themselves is given by

$$V_1(p_1; z)V_2(p_2; w) = \sum_{k=-\infty}^{\infty} (z-w)^k C_k(p_1, p_2)V_k(p_1 + p_2; \frac{z+w}{2}) \tag{4.19}$$

where $C^k$ are the OPE coefficients and $V_k$ are some operators. Then the operator product of their BRST-invariant homotopy transforms is given by

$$K_U \circ V_1(p_1;z)K_U \circ V_2(p_2;w)$$
$$= \sum_{k=-\infty}^{\infty} (z-w)^k D_k(p_1,p_2)K_U \circ W_k(p_1+p_2;\frac{z+w}{2}) \tag{4.20}$$

with the coefficients $D_k$ and operators $W_k$ defined as follows.

Let $K_U \circ V_1$ and $K_u \circ V_2$ are the transforms of $V_1$ and $V_2$ that are BRST-invariant (given the appropriate on-shell conditions on space-time fields). Then they can be represented as BRST commutators in the large space: $K_U \circ V_1 = \{Q, KW_1\}$ and $K_U \circ V_2 = \{Q, KW_2\}$ where $W_1$ and $W_2$ are (generally) some new operators in the small space (in many important cases $W_1$ and $W_2$ may actually coincide with $V_1$ and $V_2$). Let the full OPE of $W_1$ and $W_2$ be given by

$$W_1(p_1;z)W_2(p_2;w)$$
$$= \sum_{k=-\infty}^{\infty} (z-w)^k D_k(p_1,p_2)W_k(p_1+p_2;\frac{z+w}{2}) \tag{4.21}$$

with certain operators and coefficients $W_k$ and $D_k$. Then the OPE of the homotopy transforms $K_U \circ V_1(z)$ and $K_U \circ V_1(w)$ is given by (4.20). Indeed,

$$K_U \circ V_1(z)K_U \circ V_2(w) = \{Q, KW_1\}(z)\{Q, LW_2\}(w)$$
$$= \{Q, V - K[Q,V](z) : LV : (w)$$
$$= \{Q, \sum_{k=-\infty}^{\infty} K(w)D_k(p_1,p_2)W_k(p_1+p_2;\frac{z+w}{2})\}$$
$$-\{Q, K[Q,W_1](z)KW_2(w)\} \tag{4.22}$$

where we used the BRST invariance of $\{Q, KW_1\}$ and the OPE (4.21) of $W_1$ and $W_2$. The OPE (4.20) is then given by

$$K_U \circ V_1(z)K_U \circ V_2(w) =$$
$$K_U \circ (W_1(z)W_2(w)) - \{Q, L[Q,V](z)LV(w)\} \tag{4.23}$$

The first term in this OPE coincides with the right hand side of (4.20). The second term is the BRST commutator in the small Hilbert space. Indeed, if the OPEs of $K$ with $W_1$ and $W_2$ are nonsingular, one can cast the second term in (4.23) as

$$\{Q, K[Q, W_1](z)KW_2(w)\} = \{Q, C(z, w)\}$$
$$C(z, w) = \sum_{m=0}^{\infty}(z - w)^m W_1(z)[Q, \partial^m LLW_2](w) \tag{4.24}$$

Since $C(z, w)$ is the product of $W_1$ (operator in the small Hilbert space) and the BRST commutator in the large Hilbert space, it is the element of the small Hilbert space. This concludes the proof of the formula (4.20), up to BRST exact terms in the small space, irrelevant for the beta-function. The relation (4.20) is remarkably useful, since it allows us to replace the computation of the products of homotopy-transformed operators (which manifest expressions are cumbersome and complicated) with the products of operators which structure is far simpler. The sigma-model we consider is given by:

$$Z(e, \omega) = \int D[X, \psi, \bar{\psi}, ghosts]e^{-S_{RNS}+\int d^d p G(p)} \tag{4.25}$$

The leading order contributions to the $\beta$-function are given by terms quadratic in $G(p)$ and are proportional to $e^2$, $\omega^2$ and $e\omega$. Consider the contribution proportional to $e^2$ first. It is given by

$$\frac{1}{2}\int_p\int_q e^a_m(p)e^b_n(q)(F^m\bar{L}_a(p)F^n\bar{L}_b(q) + c.c.)$$
$$= \int_p\int_q e^a_m(p)e^b_n(q)\{(L^m + K \circ \int dz\lambda\psi^m)$$
$$\times\bar{L}_a(L^n + K \circ \int dw\lambda\psi^n)\bar{L}_b + c.c.\}$$
$$= -\frac{1}{\rho^2}\int_p\int_q\int\frac{d^2\xi_1}{|\xi_1|^2}e^a_m(p)e^b_n(q)(F^{mn}\bar{L}_{ab}(p+q) + c.c.)$$
$$= -\frac{1}{2\rho^2}log\Lambda\int_p\int_q e(p)\wedge e(q)(F\bar{L}(p+q) + c.c.) \tag{4.26}$$

where $\xi_1 = z_1 - z_2$, $\Lambda$ is worldsheet cutoff and we used the homotopy OPE property (4.20), as well as the fact that the operator in front of the exponent in the expression for $L^a$ (similarly for $\bar{L}^a$ has no OPE singularities with $e^{ipX}$ or $e^{iqX}$), up to BRST-exact terms. One can easily recognize this logarithmic divergence contributing the cosmological term to the low energy effective equations of motion. Similarly, the term quadratic in $\omega$ contributes to the beta-function as

$$\frac{1}{2}\int_p\int_q \omega_m^{ab}(p)\omega_n^{cd}(q)(F_a^m\bar{L}_b - \frac{1}{2}F_{ab}\bar{L}^m + c.c.)(p)$$

$$\times(F_c^n\bar{L}_d - \frac{1}{2}F_{cd}\bar{L}^n + c.c.)(q)$$

$$= \int_p\int_q\int \frac{d^2\xi_1}{|\xi_1|^2}\{\omega_{[m}^{ab}(p)\omega_{n]}^{ad}(q)(F^{mn}\bar{L}_{bd} + c.c.)(p+q)\}$$

$$= log\Lambda\int_p\int_q[\omega(p)\wedge\omega(q)(F\bar{L} + c.c.)(p+q)] \quad (4.27)$$

Thus the right-hand side (4.27) accounts for $\omega\wedge\omega$ contribution to the $\beta$-function.

Also the divergence due to cross-terms proportional to $\sim e\omega$ vanishes provided that the zero torsion constraint is satisfied.

Altogether this implies the vanishing of the conformal beta-function for the model (4.25) leads to the low-energy effective equations of motion:

$$R^{ab} = d\omega^{ab} + (\omega\wedge\omega)^{ab} - \frac{1}{\rho^2}e^a\wedge e^b = 0 \quad (4.28)$$

and

$$de^a + \omega^{ab}\wedge e^b = 0 \quad (4.29)$$

which describe the AdS gravity in MMSW formalism. The cosmological term with $\Lambda = -\frac{1}{\rho^2}$ originates from the transvection symmetry generators that serve as building blocks for the vertex operators. Thus the leading order contribution to the $\beta$-function in the sigma-model model (4.25) describes the *AdS* vacuum solution of the MMSW gravity with negative cosmological constant. As we only considered the lowest order contributions the beta-function (4.26), (4.27) we only recovered the vacuum solution with no fluctuations. The important next step will be to consider the fluctuations of higher-spin modes around the AdS vacuum. For that, one has to extend (4.25) by adding terms with vertex operators, describing the higher spin fluctuations in the frame-like approach, which we will consider next.

## 4.2 Vertex Operators for Symmetric Higher-Spin Frame-Like Fields: The Spin 3 Case

As a warm-up example, we start with the vertex operator construction for the spin 3 symmetric field in the frame-like formalism. The set of the frame-like fields describing $s = 3$ are $\Omega^{2|0}$, $\Omega^{2|1}$ and $\Omega^{2|2}$ (one dynamical and two extra fields).

In particular, in RNS superstring theory the operators for $s = 3$ are given by

$$V^{(-3)} = H_{abm}(p)ce^{-3\phi}\partial X^a \partial X^b \psi^m e^{ipX} \tag{4.30}$$

at unintegrated minimal negative picture and

$$V^{(+1)} = K \circ H_{abm}(p) \oint dz e^{\phi} \partial X^a \partial X^b \psi^m e^{ipX} \tag{4.31}$$

at integrated minimal positive picture $+1$. The operators (4.30) and (4.31) are the elements of negative and positive ghost cohomologies $H_{-3}$ and $H_1$ respectively. They are related according to $V^{(+1)} =: Z\Gamma^2 Z\Gamma^2 : V^{(-3)}$ by combination of BRST-invariant transformations by picture-changing operators for $b-c$ and $\beta-\gamma$ systems: $Z =: b\delta(T):$ and $\Gamma =: \delta(\beta)G:$ ($T$ is the full stress tensor and $G$ is the supercurrent), therefore the on-shell conditions and gauge transformations for $H_{abm}$ at positive and negative pictures are identical. The manifest expression for $V^{(+1)}$ is given by

$$\begin{aligned} V_{s=3}(p;w) &= \oint dz(z-w)^2 U(z) \\ &\equiv A_0 + A_1 + A_2 + A_3 + A_4 + A_5 + A_6 + A_7 + A_8 \end{aligned} \tag{4.32}$$

where

$$\begin{aligned} &A_0(p;w) \\ &= \frac{1}{2} H_{abm}(p) \oint dz(z-w)^2 B^{(2)}_{2\phi-2\chi-\sigma} e^{\phi} \partial X^a \partial X^b \psi^m e^{i\vec{p}\vec{X}}(z) \end{aligned} \tag{4.33}$$

and

$$A_8(w) = H_{abm}(p) \oint dz(z-w)^2 \partial cc\partial\xi\xi e^{-\phi} \partial X^a \partial X^b \psi^m \} e^{i\vec{p}\vec{X}}(z) \tag{4.34}$$

have ghost factors proportional to $e^{\phi}$ and $\partial cc\partial\xi\xi e^{-\phi}$ respectively and the rest of the terms carry ghost factor proportional to $c\xi$:

$$
\begin{aligned}
&A_1(p;w) = -2H_{abm}(p)\\
&\times \oint dz(z-w)^2 c\xi(\vec{\psi}\partial\vec{X})\partial X^a \partial X^b \psi^m e^{i\vec{p}\vec{X}}(z)\\
&A_2(p;w) = -H_{abm}(p)\\
&\times \oint dz(z-w)^2 c\xi \partial X^a \partial X^b \partial X^m P^{(1)}_{\phi-\chi} e^{i\vec{p}\vec{X}}(z)\\
&A_3(p;w) = H_{abm}(p) \oint dz(z-w)^2 c\xi \partial X^a \partial X^b \partial^2 X^m e^{i\vec{p}\vec{X}}(z)\\
&A_4(p;w) = 2H_{abm}(p) \oint dz(z-w)^2 c\xi \partial\psi^a P^{(1)}_{\phi-\chi} \partial X^b \psi^m e^{i\vec{p}\vec{X}}(z)\\
&A_5(p;w) = 2H_{abm}(p) \oint dz(z-w)^2 c\xi \partial^2\psi^a \partial X^b \psi^m e^{i\vec{p}\vec{X}}(z)\\
&A_6(p;w) = -2H_{abm}(p)\\
&\times \oint dz(z-w)^2 c\xi \partial X^a \partial X^b (\partial^2 X^m + \partial X^{a_3} P^{(1)}_{\phi-\chi}) e^{i\vec{p}\vec{X}}(z)\\
&A_7(p;w) = 2iH_{abm}(p)\\
&\times \oint dz(z-w)^2 c\xi(\vec{p}\vec{\psi}) P^{(1)}_{\phi-\chi} \partial X^a \partial X^b \psi^m e^{i\vec{p}\vec{X}}(z)\\
&A_8(p;w) = 2iH_{abm}(p) \oint dz(z-w)^2 c\xi(\vec{p}\partial\vec{\psi})\partial X^a \partial X^b \psi^m e^{i\vec{p}\vec{X}}(z)
\end{aligned}
\tag{4.35}
$$

Here $w$ is an arbitrary point in on the worldsheet; as we explained above, since all the $w$-derivatives of $s = 3$ operators are BRST-exact in small Hilbert space [86], all the correlation functions involving higher spin operators $V_{s=3}(p,w)$ are $w$-independent and the choice of $w$ is arbitrary.

As it is straightforward to check, the BRST-invariance constraints on the operators (4.30) and (4.31) lead to Pauli-Fierz type conditions

$$
p^2 H_{abm} = p^a H_{abm} = \eta^{ab} H_{abm} = 0 \tag{4.36}
$$

However, in general

$$
\eta^{am} H_{abm} \neq 0 \tag{4.37}
$$

as the tracelessness in $a$ and $m$ or $b$ and $m$ indices isn't required for $V^{(-3)}$ to be primary field. In what follows below we shall interpret $H_{abm}$ with the dynamical spin 3 connection form $\omega^{2|0}$, identifying $m$ with the manifold index and $a, b$ with the fiber indices. So the tracelessness condition is generally imposed by BRST invariance constraint on any pair of fiber indices only (but not on a pair of manifold and fiber indices). The same is actually true also for the vertex operators for frame-like gauge fields of spins higher than 3. Altogether, this corresponds precisely to the double tracelessness constraints for corresponding metric-like Fronsdal's fields for higher spins (although the zero double trace condition does not of course appear in the case of $s = 3$). As it is clear from the manifest expressions (4.30), (4.31) the tensor $H_{abm}$ is by definition symmetric in indices $a$ and $b$ and therefore can be represented as a sum of two Young diagrams. However, only the fully symmetric diagram is the physical state, since the second one (with two rows) can be represented as the BRST commutator in the small Hilbert space:

$$\begin{aligned}
&V^{(-3)} \sim \{Q, W\} \\
&W = H_{abm}(p)c\partial\xi e^{-4\phi+ipX}\partial X^a(\psi^{[m}\partial^2\psi^{b]} - 2\psi^{[m}\partial\psi^{b]}\partial\phi \\
&+\psi^m\psi^b(\frac{5}{13}\partial^2\phi + \frac{9}{13}(\partial\phi)^2)) + a \leftrightarrow b
\end{aligned} \tag{4.38}$$

If $\Omega_{abm}$ is two-row, the $V^{(-3)}$ operator is obtained as the commutator of $W$ with the matter supercurrent term of $Q$ given by $\sim \oint \gamma\psi_m\partial X^m$. As $W$ commutes with $\oint(-\frac{1}{4}b\gamma^2 - bc\partial c)$ term in $Q$, $V^{(-3)}$ is BRST-exact if and only if it commutes the stress energy part of $Q$ given by $\oint cT$. This is the case if the integrand of $W$ is a primary field. It is, however, easy to check that the integrand is primary only when the last term in its expression is present. Since this term is proportional to $\sim \partial\xi e^{-4\phi+ipX}\partial X^a\psi^m\psi^b(\frac{5}{13}\partial^2\phi + \frac{9}{13}(\partial\phi)^2)$ it is automatically antisymmetric in $m$ and $b$. and is absent when multiplied by fully symmetric $H_{abm}$. In the latter case this term is not a primary since its OPE with $T$ contains cubic singularities and therefore the commutator of $Q$ with $W$ does not give $V^{(3)}$. Similarly, shifting $H_{abm}$ by symmetrized derivative $H_{abm} \to H_{abm} + p_{(m}\Lambda_{ab)}$ is equivalent to shifting the vertex operator (7) by BRST exact terms given by

$$\begin{aligned}
&V^{(-3)} \to V^{(-3)} + \{Q, U\} \\
&U = \Lambda_{ab}c\partial\xi e^{-4\phi+ipX}\{\partial X^a((p\psi)\partial^2\psi^b - 2(p\psi)\partial\psi^b\partial\phi \\
&+(p\psi)\psi^b(\frac{5}{13}\partial^2\phi + \frac{9}{13}(\partial\phi)^2)) \\
&+\partial X^a\partial X^b((p\partial^2 X) - \partial\phi(p\partial X))\}
\end{aligned} \tag{4.39}$$

Of course everything described above also applies to the vertex operator (4.31) at positive picture, with appropriate $Z, \Gamma$ transformations. This altogether already sends a strong hint to relate (4.30), (4.31) to vertex operators for the dynamical frame-like field $\omega^{2|0}$ describing spin 3. However, to make the relation between string theory and frame-like formalism, we still need the vertex operators for the remaining extra fields $\omega^{2|1}$ and $\omega^{2|2}$. The expressions that we propose are given by

$$V^{2|1}(p) = 2\omega_m^{ab|c}(p)ce^{-4\phi}(-2\partial\psi^m\psi_c\partial X_{(a}\partial^2X_{b)} \\ -2\partial\psi^m\partial\psi_c\partial X_a\partial X_b + \psi^m\partial^2\psi_c\partial X_a\partial X_b)e^{ipX} \tag{4.40}$$

for $\omega^{2|1}$ and

$$V^{2|2}(p) = -3\omega_m^{ab|cd}(p)ce^{-5\phi}(\psi^m\partial^2\psi_c\partial^3\psi_d\partial X^a\partial X_b \\ -2\psi^m\partial\psi_c\partial^3\psi_d\partial X_a\partial^2X_b \\ +\frac{5}{8}\psi^m\partial\psi_c\partial^2\psi_d\partial X_a\partial^3X_b + \frac{57}{16}\psi^m\partial\psi_c\partial^2\psi_d\partial^2X_a\partial^2X_b)e^{ipX} \tag{4.41}$$

for $\omega^{2|2}$. We start with analyzing the operator for $\omega^{2|1}$. Straightforward application of $\Gamma$ to this operator gives

$$:\Gamma V^{2|1}:(p) = V^{(-3)}(p) \\ H_m^{ab}(p) = ip_c\omega_m^{ab|c}(p) \tag{4.42}$$

i.e. the picture-changing of $V^{2|1}$ gives the vertex operator for $\omega^{2|0}$ with the 3-tensor given by the divergence of $\omega^{2|1}$, i.e. for $p_c\omega_m^{ab|c}(p) \neq 0$ $V^{2|1}$ is the element of $H_{-3}$. If, however, the divergence vanishes, the cohomology rank changes and $V^{2|1}$ shifts to $H_{-4}$. This is precisely the case we are interested in. Namely, consider the $H_{-4}$ cohomology condition

$$p_c\omega_m^{ab|c}(p) = 0 \tag{4.43}$$

The general solution of this constraint is

$$\omega_m^{ab|c} = 2p^c\omega_m^{ab} - p^a\omega_m^{bc} - p^b\omega_m^{ac} + p_d\omega_m^{acd;b} \tag{4.44}$$

where $\omega_m^{ab}$ is traceless and divergence free in $a$ and $b$ and satisfies the same on-shell constraints as $H_m^{ab}$, while $\omega_m^{acd;b}$ is some three-row field, antisymmetric in $a, c, d$ and symmetric in $a$ and $b$. It is, however, straightforward to check that the operator $V^{2|1}$ with the polarization given by $\omega^{ab|c} = p_d\omega_m^{acd;b}$ can be cast as the BRST commutator:

$$p_d\omega_m^{acd;b}(p)V^m_{ac|b}(p) \\ = \{Q, \omega_m^{acd;b}(p)\oint dz e^{\chi-5\phi+ipX}\partial\chi(-2\partial\psi^m\psi_c\partial X_a\partial^2X_b \\ -2\partial\psi^m\partial\psi_c\partial X_a\partial X_b + \psi^m\partial^2\psi_c\partial X_a\partial X_b) \\ \times(\partial^2\psi_d - \frac{4}{3}\partial\psi_d\partial\phi + \frac{1}{141}\psi_d(41(\partial\phi)^2 - 29\partial^2\phi))\} \tag{4.45}$$

therefore, modulo pure gauge terms the cohomology condition (4.43) is the zero torsion condition relating the extra field $\omega^{2|1}$ to the dynamical $\omega^{2|0}$ connection. Similarly, constraining $V^{2|2}$ to be the element of $H_{-5}$ cohomology results in the second generalized zero torsion condition

$$\omega_m^{ab|cd} = 2p^d\omega^{ab|c} - p^a\omega^{bd|c} - p^b\omega^{ad|c}$$
$$+2p^c\omega^{ab|d} - p^a\omega^{bc|d} - p^b\omega^{ac|d} \tag{4.46}$$

relating $\omega^{2|2}$ to $\omega^{2|1}$ modulo BRST-exact terms $\sim \{Q, W^{2|2}(p)\}$ where

$$W^{2|2}(p) = \omega^{ab;cdf}(p)\oint dze^{ipX}[(\psi^m\partial^2\psi_c\partial^3\psi_d\partial X^a\partial X_b$$
$$-2\psi^m\partial\psi_c\partial^3\psi_d\partial X_a\partial^2 X_b$$
$$+\frac{5}{8}\psi^{(m}\partial\psi_c\partial^2\psi_{d)}\partial X_a\partial^3 X_b + \frac{57}{16}\psi^m\partial\psi_c\partial^2\psi_d\partial^2 X_a\partial^2 X_b)]$$
$$\times(-\frac{5}{2}L_f\partial^2\xi + \partial L_f\partial\xi) \tag{4.47}$$

where, as previously, $\xi = e^\chi$ and

$$L_f = e^{-6\phi}(\partial^2\psi_f - \partial\psi_f\partial\phi + \frac{3}{25}\psi_f((\partial\phi)^2 - 4\partial^2\phi)) \tag{4.48}$$

The gauge transformations for $\omega^{2|1}$ and $\omega^{2|2}$: $\delta\omega_m^{ab|c}(p) = p_m\Lambda^{ab|c}$ and $\delta\omega_m^{ab|cd}(p) = p_m\Lambda^{ab|cd}$ with $\Lambda$'s having having the same symmetries in the fiber indices as $\omega$'s shift the vertex operators by terms that are BRST-exact in the small Hilbert space; the explicit expressions for the appropriate BRST commutators are given in the Appendix B. Similarly to the $\omega^{2|0}$ case, for $\omega^{2|1}$ and $\omega^{2|2}$ with the manifold $m$ index antisymmetric with any of the fiber indices $a$ or $b$ the operators (4.30), (4.41) become BRST-exact in the small Hilbert space. Given the cohomology ("zero torsion") condition (4.43), this ensures that the fully symmetric symmetric $s = 3$ Fronsdal field is related to dynamical field $\omega^{2|0}$ by the gauge transformation removing the two-row diagram.

This concludes the construction of the vertex operators for frame-like gauge fields for spin 3. In the next section we shall extend this construction for the higher spin values, and analyze the on-shell conditions on space-time gauge fields in the leading order, demonstrating the emergence of0 the *AdS* geometry.

## 4.3 Higher Spins Operators, Weyl Invariance and Emergence of AdS Geometry

Given the global symmetry generators, it is straightforward to construct the appropriate vertex operators in open and closed string theories describing emissions of massless particles of various spins. As the generators of nonzero ghost cohomologies in "larger" string theory induce isometries in *AdS* space and give rise to enveloppings of these isometries (AdS higher-spin algebras), the next step is to point out how the AdS geometry emerges in the $\beta$-function equations for the massless higher spin modes in RNS theory. The $\beta$-function equations can be obtained from the scale invariance (Weyl invariance) constraints in string perturbation theory.

As we shall find out, in the leading order, Weyl invariance constraints on the higher spin vertex operators lead to low-energy equations of motion for massless higher spin fields defined by the Fronsdal operator in AdS space-time. The AdS structure of the Fronsdal operator (with the appropriate mass-like terms) emerges despite the fact that the higher spin vertex operators are initially defined in the flat background in RNS theory. The appearance of the AdS geometry is directly related to the ghost cohomology structure of the higher spin vertices and is detected through the off-shell analysis of the $2d$ scale invariance of the vertex operators for higher spins. It is crucial that, in order to see the emergent *AdS* geometry one must go off-shell, e.g. to analyze the scale invariance of the operators in $2+\epsilon$ dimensions so that the trace $T_{z\bar{z}}$ of the stress-energy tensor generating $2d$ Weyl transformations is no longer identically zero. Namely, it is the off-shell analysis of the operators at nonzero $H_n$ that allows to catch cosmological type terms in low energy equations of motion while the on-shell constraints on the operators (such as BRST conditions) do not detect them, only leading to standard Pauli-Fierz equations for massless higher spins in flat space. This is a strong hint that the, from the string-theoretic point of view, the appropriate framework to analyze the higher spin interactions is the off-shell theory, i.e. string field theory, with the SFT equations of motion: $Q\Psi = \Psi \star \Psi$ related to Vasiliev's equations in unfolding formalism. It is important to stress, however, that Vasiliev's equations must be related to the enlarged, rather than ordinary SFT, with the string field $\Psi$ extended to higher ghost cohomologies. Higher spin interactions in AdS should then be deduced from the off-shell string field theory computations involving higher spin vertex operators for Vasiliev's frame-like fields on the worldsheet boundary, with the appropriate insertions of $T_{z\bar{z}}$ in the bulk controlling the cosmological constant dependence. We will address this question below. We start with the spin 2 case. Using the *AdS* transvection operators (3.31)-(3.33) and the analogy with the usual string graviton (constructed as the bilinear in Poincare translations), consider the closed string spin 2 operator in $H_{-3} \otimes H_{-3}$ cohomology:

$$
\begin{aligned}
&V_{s=2} = G_{mn}(p) \int d^2 z e^{-3\phi - 3\bar{\phi}} R^m \bar{R}^n e^{ipX}(z, \bar{z}) \\
&R^m = \bar{\lambda}\partial^2 X^m - 2\partial\lambda\partial X^m \\
&+ip^m(\frac{1}{2}\partial^2\lambda + \frac{1}{q}\partial\varphi\partial\lambda - \frac{1}{2}\lambda(\partial\varphi)^2 \\
&+(1+3q^2)\lambda(3\partial\psi_p\psi^p - \frac{1}{2q}\partial^2\varphi))\} \\
&m = 0, ..., d-1
\end{aligned} \tag{4.49}
$$

where $G_{mn}$ is symmetric. The operator on $H_1 \otimes H_1$ is constructed likewise by replacing $\int d^2 z \to K\bar{K} \circ \int d^2 z$ and $-3\phi \to \phi, -3\bar{\phi} \to \bar{\phi}$. Provided that $k^2 = 0$,0 it is straightforward to check its BRST-invariance with respect to the flat space BRST operator as well as the linearized diffeomorphism invariance since the transformation $G^{mn}(p) \to G^{mn}(p) + p^{(m}\epsilon^{n)}$ shifts holomorphic and antiholomorphic factors of $V_{s=2}$ by terms BRST-exact in small Hilbert space. The operator (4.49) appears a natural candidate to describe gravitational fluctuations around the $AdS$ vacuum in the *metric* formalism. In order to identify this symmetric massles spin 2 state with gravitational fluctuations, however, one needs to analyze the low-energy equations of motion for $G_{mn}$ which leading order is given by the Weyl constraints. So we shall study the invariance constraints on the operator (4.49) of $H_{-3} \otimes H_{-3}$ (a candidate for the AdS graviton), comparing them to those for the ordinary graviton in RNS theory. To see the difference, let us first recall the most elementary example - the graviton in bosonic string theory given by

$$
V = G_{mn} \int d^2 z \partial X^m \bar{\partial} X^n e^{ipX} \tag{4.50}
$$

The condition $[Q, V] = 0$ leads to constraints : $p^2 G_{mn}(p) = p^m G_{mn}(p) = 0$ related to linearized Ricci tensor contributions to the graviton's $\beta$-function. The complete linearized contribution to the graviton's $\beta$-function, however, is given by $\beta_{mn} = R_{mn}^{linearized} + 2\partial_m\partial_n D$ (where $D$ is the space-time dilaton) with the last term particularly accounting for the $\sim e^{-2D}$ factor in the low-energy effective action. This term in fact is *not* produced by any of the on-shell (BRST) constraints on the graviton vertex operator; to recover it, one has to analyze the $off-shell$ constraints related to the Weyl invariance. Namely, the generator of the Weyl transformations is given by, the $T_{z\bar{z}}$ component of the stress-energy which is identically zero on-shell in $d = 2$ but nonzero in $d = 2+\epsilon$. The leading order contribution to the $\beta$-function of closed string vertex operator $V$ is determined by the coefficient in front of $\sim \frac{1}{|z-w|^2}$ in the midpoint OPE of $T_{z\bar{z}}(z,\bar{z})$ and $V(w,\bar{w})$ leading to logarithmic divergence in the integral $\sim \int d^2z \int d^2w T_{z\bar{z}}(z,\bar{z})V(w,\bar{w})$. In bosonic string theory one has

$$T_{z\bar{z}} \sim -\partial X_m \bar{\partial} X^m + \partial\sigma\bar{\partial}\sigma + \partial\bar{\partial}(...) \tag{4.51}$$

skipping the full-derivative part proportional to $2d$ Laplacian related to background charge, as it leads to contact terms in the OPE with V, not contributing to its $\beta$-function. We then easily calculate:

$$\begin{aligned}
&\int d^2z \int d^2w T_{z\bar{z}}(z,\bar{z})V(w,\bar{w}) \\
&\sim G_{mn}(p)\int d^2z \int d^2w \frac{1}{|z-w|^2} e^{ipX}(\frac{z+w}{2},\frac{\bar{z}+\bar{w}}{2}) \\
&\times\{p^2\partial X^m\bar{\partial}X^n - \frac{1}{2}(p^m p_s \partial X^s \bar{\partial} X^n + p^n p_s \partial X^m \bar{\partial} X^s) \\
&+\frac{1}{4}\eta^{mn}p_s p_t \partial X^s \bar{\partial} X^t\}(\frac{z+w}{2},\frac{\bar{z}+\bar{w}}{2}) \\
&\sim ln\Lambda[p^2 G_{mn}(p) - \frac{1}{2}(p^s p_m G_{ns}(p) + p^s p_n G_{ms}(p)) \\
&+2p_m p_n D(p)]\int d^2\zeta \partial X^m \bar{\partial} X^n e^{ipX}(\zeta,\bar{\zeta})
\end{aligned} \tag{4.52}$$

where we introduced $ln\Lambda = \int \frac{d^2\xi}{|\xi|^2}, \zeta = z+w$ and identified the dilaton with the trace of the space-time metric: $D(p) \sim \eta^{st}G_{st}(p)$.

For conventional reasons and in order not to introduce too many letters, we adopt the same notation, $\Lambda$, for both the worldsheet cutoff and the cosmological constant in space time. However, we hope that the distinction between those will be very clear to a reader from the context; in particular, the cutoff shall always appears in terms of logs, while all expressions in the cosmological constants are either linear or polynomial.

The coefficient in front of the integral thus determines the leading order contribution to the graviton's $\beta$-function. The first three terms in this coefficient simply give linearized Ricci tensor while the last one proportional to the trace of the space-time metric determines the string coupling dependence. All these terms contain two space-time derivatives and obviously no cosmological-type contributions appear. The analogous calculation is of course similar in RNS superstring theory, producing the similar answer. However, in comparison with the bosonic string the RNS case also contains some instructive subtlety which will be useful to observe for future calculations. That is, consider graviton operator in RNS theory at canonical $(-1,-1)$-picture:

$$V^{(-1,-1)} = G_{mn}(p) \int d^2 z e^{-\phi-\bar{\phi}} \psi^m \bar{\psi}^n e^{ipX}(z,\bar{z}) \quad (4.53)$$

The generator of Weyl transformations in RNS theory is

$$\begin{aligned} T_{z\bar{z}}^{RNS} &= T_X + T_\psi + T_{b-c} + T_{\beta-\gamma} + T_{Liouv} \\ &= -\frac{1}{2}\partial X_m \bar{\partial} X^m - \frac{1}{2}(\bar{\partial}\psi_m \psi^m + \partial\bar{\psi}_m \bar{\psi}^m) \\ &+ \frac{1}{2}\partial\sigma\bar{\partial}\bar{\sigma} - \frac{1}{2}\partial\phi\bar{\partial}\bar{\varphi} + \frac{1}{2}\partial\chi\bar{\partial}\bar{\chi} + \partial\bar{\partial}(...) \end{aligned} \quad (4.54)$$

The contribution of $T_X$ to the scale transformation is again easily computed to give $\sim p^2 G_{mn}(p) ln\Lambda V^{(-1,-1)}$, i.e. the gauge fixed linearized Ricci tensor (with the gauge condition $p^m G_{mn} = 0$ imposed by invariance under transformations of $V^{(-1,-1)}$ by worldsheet superpartners of $T_X$, namely, $G_{+\bar{z}}$ and $G_{-z}$). To compute the contribution fom $T_\psi$, it is convenient to bosonize $\psi$ according to

$$\begin{aligned} &\psi_1 \pm i\psi_2 = e^{\pm i\varphi_1} ... \\ &\psi_{d-1} \pm i\psi_d = e^{\pm i\varphi_{\frac{d}{2}}} \end{aligned} \quad (4.55)$$

(for simplicity we can assume the number $d$ of dimensions even, without loss of generality). Then the stress-energy tensor for $\psi$ is

$$T = -\frac{1}{2}(\bar{\partial}\psi_m \psi^m + \partial\bar{\psi}_m \bar{\psi}^m) = \sum_{i=1}^{\frac{d}{2}} \partial\varphi_i \bar{\partial}\bar{\varphi}^i \quad (4.56)$$

Writing $\psi_1 = \frac{1}{2}(e^{i\varphi_1} - e^{-i\varphi_1})$, it is easy to compute the contribution of $T_\psi$ to the $\beta$-function:

$$\begin{aligned} &\int d^2 z T_\psi(z,\bar{z}) G_{mn}(p) \int d^2 w e^{-\phi-\bar{\phi}} \psi^m \bar{\psi}^n e^{ipX}(w,\bar{w}) \\ &= \frac{1}{2} ln\Lambda G_{mn}(p) \int d^2\zeta e^{-\phi-\bar{\phi}} \psi^m \bar{\psi}^n e^{ipX}(\zeta,\bar{\zeta}) \equiv \frac{1}{2} ln\Lambda V^{(-1,-1)} \end{aligned} \quad (4.57)$$

Note the scale transformation by $T_\psi$ contributes the term proportional to $\sim \frac{1}{2} G_{mn}$ with no derivatives, i.e. a “cosmological” type term. The cosmological term is of course absent in the overall graviton’s $\beta$-function as the contribution (4.57) is precisely cancelled by the scale transformation of the ghost part of $V^{(-1,-1)}$ by $T_{\beta\gamma} = -\frac{1}{2}|\partial\phi|^2 + \partial\bar{\partial}(...)$:

$$\int d^2 z T_{\beta\gamma}(z,\bar{z}) G_{mn}(p) \int d^2 w e^{-\phi-\bar{\phi}} \psi^m \bar{\psi}^n e^{ipX}(w,\bar{w})$$
$$= -\frac{1}{2} ln \Lambda V^{(-1,-1)} \tag{4.58}$$

with the minus sign related to that of the $\phi$-ghost field in the trace of the stress-energy tensor. So the absence of the cosmological term in the graviton's $\beta$-function in RNS theory is in fact the result of the smart cancellation between the Weyl transformations of the matter and the ghost factors of the graviton operator at $(-1,-1)$ canonical picture (despite that the final answer - the absence of the overall cosmological term may seem obvious). The same result of course applies to the flat graviton operator transformed to any other ghost picture since it is straightforward to check that both $\Gamma$ and $\Gamma^{-1}$ are Weyl-invariant, up to BRST-exact terms. The absence of cosmological (or mass-like) terms in the Weyl transformation laws is actually typical for any massless operators of $H_0$ or $H_0 \otimes H_0$; at nonzero pictures it is the consequence of the cancellation of Weyl transformations for the matter and the ghosts, as was demonstrated above. This observation is of importance since, as it will be shown below, this matter-ghost cancellation does *not* occur for operators of nonzero $H_n$'s, in particular, for the spin 2 operator (4.49). In closed string theory and for massless operators for higher spin fields of Vasiliev type in open string sector. Namely, we will show that for the operator (4.49) the scale invariance constraints lead to cosmological term, while for massless higher spin fields the similar constraints lead to emergence of AdS geometry in the Fronsdal's operator in the low-energy limit. We start with analyzing the scale transformation of the spin 2 operator (4.49) by $T_X$. The canonical picture for the operator (4.49) is $(-3,-3)$. To deduce the transformation law for the operator (4.49) it is sufficient to consider the momentum-independent part $\sim R_0^m \bar{R}_0^n$ of the the matter factor $\sim R^m \bar{R}^n$ in (4.49):

$$R^m = R_0^m + ik^m(...)$$
$$R_0^m = \lambda\partial^2 X^m - 2\partial\lambda\partial X^m \tag{4.59}$$

and similarly for $\bar{R}^m$ Then the straightforward application of $T_X$ to

$$G_{mn}(p) \int d^2 w e^{-3\phi-3\bar{\phi}} R_0^m \bar{R}_0^n e^{ipX}(w,\bar{w})$$

gives

$$\int d^2 z T_X(z,\bar{z}) G_{mn}(p) \int d^2 w e^{-3\phi-3\bar{\phi}} R_0^m \bar{R}_0^n e^{ipX}(w,\bar{w})$$
$$= ln\Lambda \times G_{mn}(p) \int d^2\zeta \{ -\frac{1}{2} p^2 e^{-3\phi-3\bar{\phi}} R_0^m \bar{R}_0^n e^{ipX}(\zeta,\bar{\zeta})]$$
$$-\frac{i}{8} p^m \partial^2 (e^{-3\phi}\lambda e^{ipX}(\zeta)) e^{-3\bar{\phi}} \bar{R}_0^n e^{ipX}(\bar{\zeta})$$
$$+\frac{i}{2} p^m \partial\partial (e^{-3\phi}\partial\lambda e^{ipX}(\zeta)) e^{-3\bar{\phi}} \bar{R}_0^n e^{ipX}(\bar{\zeta}) + (c.c.; m \leftrightarrow n)]\}$$
$$= ln\Lambda G_{mn}(p) \int d^2\zeta (-\frac{1}{2} p^2 \delta_q^n + \frac{1}{2} p^m p_q) e^{-3\phi-3\bar{\phi}} R_0^m \bar{R}_0^q e^{ipX} + ... \tag{4.60}$$

where we dropped BRST-exact terms and only kept terms contributing to the $G_{mn}$'s $\beta$-function, skipping those relevant to $\beta$-functions of the space-time fields other than $G_{mn}$. In addition, for simplicity we skipped the dilaton-type contributions involving the trace of $G_{mn}$; it is, however, straightforward to generalize the computation to include the dilaton, accounting for the standard factor of $e^{-2D}$ in the effective action. Comparing the transformation laws (4.52) and (4.60), we conclude that the contribution of $T_X$-transformation to the $G_{mn}$ $\beta$-function results in the linearized Ricci tensor $R_{mn}^{linearized}$. Next, consider the contributions from $T_\lambda = -\frac{1}{2}(\partial\bar{\lambda}\bar{\lambda} + \bar{\partial}\lambda\lambda)$ and $T_{\beta\gamma}$ to $\beta_{mn}$. The analysis is similar to the one for the ordinary graviton operator, however, the crucial difference is that this time there is no cancellation between transformations due to the worldsheet matter (Liouville) fermion and the $\beta-\gamma$ ghost, observed above. As previously, the transformation of by $T_\lambda$ contributes

$$G_{mn}(p) \int d^2 z T_\lambda(z,\bar{z}) \int d^2 w e^{-3\phi-3\bar{\phi}} R_0^m \bar{R}_0^n e^{ipX}(w,\bar{w})$$
$$= \frac{1}{2} ln\Lambda G_{mn}(p) \int d^2\zeta e^{-3\phi-3\bar{\phi}} R_0^m \bar{R}_0^n e^{ipX}(w,\bar{w}) \tag{4.61}$$

On the other hand, the transformation by $T_{\beta-\gamma}$ produces:

$$G_{mn}(p) \int d^2 z T_{\beta-\gamma}(z,\bar{z}) \int d^2 w e^{-3\phi-3\bar{\phi}} R_0^m \bar{R}_0^n e^{ipX}(w,\bar{w})$$
$$= -\frac{9}{2} ln\Lambda G_{mn}(p) \int d^2\zeta e^{-3\phi-3\bar{\phi}} R_0^m \bar{R}_0^n e^{ipX}(w,\bar{w}) \tag{4.62}$$

where we used the OPE $|\partial\phi|^2)z,\bar{z}e^{-3\phi-3\bar{\phi}}(w,\bar{w}) \sim \frac{9}{|z-w|^2} e^{-3\phi-3\bar{\phi}}(w,\bar{w})$ Unlike the case of the ordinary graviton, the cosmological type contributions from the scale transformations of the ghost and the matter part of the operator (4.49) no longer cancel each other. As a result, the overall cosmological term $\sim (\frac{9}{2} - \frac{1}{2})G_{mn}$ appears in the $\beta$-function of the spin 2 operator, which leading order is now given by

$$\beta_{mn} = R_{mn}^{linearized} - 8G_{mn} \tag{4.63}$$

(with the extra factor of 2 related to the normalization of the Ricci tensor). The appearance of the cosmological term is thus closely related to the ghost cohomology structure of the operator (4.49), i.e. to the fact that the canonical picture for this operator is $(-3, -3)$ while the standard $(-1, -1)$ picture representation of the "ordinary" graviton does not exist for (4.49). This altogether allows us to identify the space-time massless spin 2 $G_{mn}$ field emitted by $H_{-3} \otimes H_{-3}$ with the gravitational fluctuations around the $AdS$ vacuum. As was pointed out above, this is not a surprise since the operator (4.49) has been originally built as a bilinear of the generators realizing transvections in $AdS$. The next step is to generalize the above arguments to the vertex operators for the massless higher spin fields (with $s \geq 3$) which are also the elements of nonzero cohomologies $H_{s-2} \sim H_{-s}$. In analogy with the mechanism generating the cosmological term in the $\beta$-function of the spin 2 operator (4.49), we expect that the scale invariance analysis of these operators shall also lead to appearance of the mass-like terms in their $\beta$-functions (although the operators by themselves are massless). We shall attempt to show that the "mass-like" terms are in fact related to the $AdS$ geometry couplings of the higher spin fields, adding up to appropriate $AdS$ Fronsdal operators in their low-energy equations of motion in the leading order.

## 4.4 Higher Spin Operators: Weyl Invariance and $\beta$-Functions

In this section we extend the analysis of the previous sections to vertex operators describing massless higher spin excitations in open RNS string theory. The space-time fields emitted by these operators correspond to symmetric higher spin gauge fields in Vasiliev's frame-like formalism. The main result of this section is that the leading order of the $\beta$-function for the higher spin operators gives the low-energy equations of motion determined by Fronsdal operator in the AdS space, despite the fact that the operators are initially defined around the flat background. As in the case of the AdS graviton considered in the previous section, the information about the AdS geometry is encrypted in the ghost cohomology structure of the operators. In the frame-like formalism, as we have seen before, a symmetric higher spin gauge field of spin $s$ is described by collection of two-row fields $\Omega^{s-1|t} \equiv \Omega_m^{a_1 \ldots a_{s-1}|b_1 \ldots b_t}(x)$ with $0 \leq t \leq s-1$ and the rows of lengths $s-1$ and $t$. The only truly dynamical field of those is $\Omega^{s-1|0}$ while the fields with $t \neq 0$, called the extra fields, are related to the dynamical one through generalized zero torsion constraints:

$$\Omega^{s-1|t} \sim D^{(t)} \Omega^{s-1|0} \tag{4.64}$$

where $D^{(t)}$ is certain order $t$ linear differential operator preserving the symmetries of the appropriate Yang tableaux. There are altogether $s-1$ constraints for the field of spin $s$. As for the dynamical $\Omega^{s-1|0}$-field (symmetric in all the $a$-indices), it splits into two diagrams with respect to the manifold $m$-index. Assuming the appropriate pullbacks, the one-row symmetric diagram describes the dynamics of the $metric-like$ symmetric Fronsdal's field of spin $s$ while the two-row component of $\Omega^{s-1|0}$ can be removed by appropriate gauge transformation. In the language of string theory, the higher spin $s$ operators are the elements of $H_{s-2} \sim H_{-s}$. The on-shell (Pauli-Fierz type) constraints on these space-time fields follow from the BRST-invariance constraints on the vertex operators, while the gauge transformations correspond to shifting the vertex operators by BRST-exact terms. The zero torsion constraints relating $\Omega^{s-1|t}$ gauge fields with different $t$ follow from the cohomology constraints on their vertex operators $V_{s-1|t}$, that is, by requiring that all these vertex operators belong to the same cohomology $H_{s-2} \sim H_{-s}$ (there are, however, certain subtleties with this scheme arising at $t=s-1$ or $t=s-2$ which were discussed above for the $s=3$ case). As in the $s=3$ case, the on-shell constraints on these space-time fields follow from the BRST-invariance constraints on the vertex operators, while the gauge transformations correspond to shifting the vertex operators by BRST-exact terms . Finally, the zero torsion constraints relating $\Omega^{s-1|t}$ gauge fields with different $t$ follow from the cohomology constraints on their vertex operators $V_{s-1|t}$, that is, by requiring that all these vertex operators belong to the same cohomology $H_{s-2} \sim H_{-s}$.

To understand the meaning of the cohomology constraints it is useful to recall first a much simpler example known from the conventional Ramond-Ramond sector of closed superstring theory. Namely, the relation between cohomology and zero torsion constraints can be thought of as a symmetric higher spin generalization of a more elementary and familiar example of standard Ramond-Ramond vertex operators in closed critical superstring theory. It is well-known that the canonical picture representation for the

Ramond-Ramond operators is given by:

$$V_{RR}^{(-\frac{1}{2},-\frac{1}{2})} = \not{F}_{\alpha\beta}(p)\int d^2z e^{-\frac{\phi}{2}-\frac{\bar{\phi}}{2}}\Sigma^\alpha\bar{\Sigma}^\beta e^{ipX}(z,\bar{z})$$
$$\not{F}^{(p)}_{\alpha\beta} \equiv \gamma_{\alpha\beta}^{m_1\dots m_p}F_{m_1\dots m_p} \quad (4.65)$$

where $\not{F}^{(p)}_{\alpha\beta}$ is the Ramond-Ramond $p$-form field strength (contracted with $10d$ gamma-matrices). Note that, since this operator is the source of the field strength ( the derivative of the gauge potential), it does not carry RR charge (which instead is carried by a corresponding Dp-brane). This Ramond-Ramond operator exists at all the half-integer pictures and is the element of $H^{(-\frac{1}{2},-\frac{1}{2})}$ cohomology (which is the superpartner of $H^{(0,0)}$ consisting of all picture-independent physical states). It is, however, possible to construct vertex operator which couples to Ramond-Ramond gauge potential rather than field strength. The canonical picture for such an operator is $(-\frac{3}{2},-\frac{1}{2})$ (or equivalently $(-\frac{1}{2},-\frac{3}{2})$ with the explicit expression given by

$$U_{RR}^{(-\frac{1}{2},-\frac{3}{2})} = \not{A}^{(p-1)}_{\alpha\beta}\int d^2z e^{-\frac{\phi}{2}-\frac{3\bar{\phi}}{2}}\Sigma^\alpha\bar{\Sigma}^\beta e^{ipX}(z,\bar{z})$$
$$\not{A}^{(p-1)}_{\alpha\beta} \equiv \gamma_{\alpha\beta}^{m_1\dots m_p}A_{m_1\dots m_{p-1}} \quad (4.66)$$

where generically, $\not{A}$ is arbitrary. The $U$-operator is generally not the picture-changed version of the $V$-operator nor it is the element of $H^{(-\frac{1}{2},-\frac{1}{2})}$ for general $\not{A}$. To relate $U_{RR}^{(-\frac{1}{2},-\frac{3}{2})}$ to $V_{RR}^{(-\frac{1}{2},-\frac{1}{2})}$ by the picture-changing:

$$V_{RR}^{(-\frac{1}{2},-\frac{1}{2})} =: \Gamma U_{RR}^{(-\frac{3}{2},-\frac{1}{2})} : \quad (4.67)$$

one has to impose the constraint $\not{F} = d\not{A}$ that ensures that $U$ is the physical operator of $H^{(-\frac{1}{2},-\frac{1}{2})}$. Thus the cohomology constraint in $U$ leads to the standard relation between the gauge potential and the field strength. Similarly, the generalized $H_{s-2} \sim H_{-s}$-cohomology constraints on higher spin operators $V_{s-1|t}$ for $\Omega^{s-1|t}$ space-time fields lead to generalized zero torsion constraints. Note that for $0 \leq t \leq s-3$ the canonical pictures for $V_{s|t}$ are $2s-t-5 \sim t+3-2s$ with the cohomology constraints $V_{s|t}{\in}H_{s-2} \sim H_{-s}$ inducing the entire chain of zero torsion relations.

With are now prepared to analalyze the scale invariance constraints for open string vertex operators describing the Vasiliev type higher spin fields in space-time. It turns out that for massless fields of spin $s$ the canonical picture representation is especially simple for the field with $t = s-3$, that is, for $\Omega^{s-1|s-3}$. The explicit vertex operator expression for this field is given by:

$$V_{s-1|s-3} = \Omega_m^{a_1...a_{s-1}|b_1...b_{s-3}}(p) \int d^2 z e^{-s\phi}$$
$$\times \psi^m \partial\psi_{b_1} \partial^2 \psi_{b_2} ... \partial^{s-3} \psi_{b_{s-3}} \partial X_{a_1} ... \partial X_{s-1} e^{ipX}$$
$$\sim \Omega_m^{a_1...a_{s-1}|b_1...b_{s-3}}(p) K \circ \int d^2 z e^{(s-2)\phi}$$
$$\times \psi^m \partial\psi_{b_1} \partial^2 \psi_{b_2} ... \partial^{s-3} \psi_{b_{s-3}} \partial X_{a_1} ... \partial X_{s-1} e^{ipX} \quad (4.68)$$

For s=3, this immediately gives the operator for the Fronsdal field. The on-shell conditions on $\Omega^{s-1|s-3}$ that ensure the BRST-invariance of $V_{s-1|s-3}$ are not difficult to obtain. The commutation with the $T_X$ component of the stress-energy part of $Q_{brst}$ leads to the tracelessness of $\Omega$ in the $a$-indices, that is, $\Omega_{ma}^{aa_1...a_{s-3}|b_1...b_{s-3}} = 0$ which is the well-known constraint on frame-like fields and to the second Pauli-Fierz constraint of transversality: $p_a \Omega_m^{aa_1...a_{s-2}|b_1...b_{s-3}}(p) = 0$. The commutation with $T_\psi$ part of $Q_{brst}$, given by $-\frac{1}{2}\oint dz c \partial\psi_p \psi^m$, requires the symmetry of $\Omega$ in the $b$-indices, as it is easy to see from the OPE between $T_\psi$ and $S_{mb_1...b_{s-3}} = \psi_m \partial\psi_{b_1} ... \partial^{s-3}\psi_{b_{s-3}}$ - the latter is the primary field of dimension $h_\psi = \frac{1}{2}(s-2)^2$ only if $S$ is symmetric and traceless in all indices. While the symmetry in the $b$-indices is another standard familiar constraint in the frame-like formalism, the symmetry and tracelessness of $m$ with respect to the $b$-indices is an extra condition on $\Omega$ that can be obtained partial fixing of the gauge symmetries of $\Omega$. Given the above conditions are fulfilled, the commutation with the supercurrent part of $Q_{brst}$ produces no new constraints, however, there is one more condition coming from the $H_{-s}$-cohomology constraint on $V_{s-1|s-3}$, that is,

$$: \Gamma V_{s-1|s-3} := 0 \quad (4.69)$$

This constraint further requires the vanishing of the mixed trace over any pair of $(a, b)$=indices: $\eta_{ab} \Omega_m^{aa_1...a_{s-2}|bb_1...b_{s-4}} = 0$. Fortunately the gauge symmetry of $\Omega$ is more than powerful enough to absorb this extra constraint as well. Finally, we are left to consider the BRST nontriviality conditions on $V_{s-1|s-3}$. First of all, the nontriviality constraint: $V_{s-1|s-3} \neq \{Q_{brst}, W_{s-1|s-3}\}$ where $W$ is some operator in small Hilbert space requires either

$$\eta_a^m \Omega_m^{aa_1...a_{s-2}|b_1...b_{s-3}} \neq 0 \quad (4.70)$$

or

$$p^m \Omega_m^{a_1...a_{s-1}|b_1...b_{s-3}} \neq 0 \quad (4.71)$$

since otherwise, generically, there exist operators

$$W_{s-1|s-3} \sim \Omega_m^{a_1...a_{s-1}|b_1...b_{s-3}} \sum_{k=0}^{s-1} \oint dz e^{\chi-(s-1)\phi} \partial\chi$$
$$\times \partial\psi_{b_1}...\partial^{s-3}\psi_{b_{s-3}} \partial X_{a_1}...\partial X_{s-1} \partial^{s-1-k} X_m G^{(k)}(\phi,\chi) e^{ipX} \quad (4.72)$$

commuting with the stress tensor part of $Q_{brst}$ while, at the same time, the commutators of the supercurrent part of $Q_{brst}$ with $W_{s-1|s-3}$ are proportional to $V_{s-1|s-3}$: $\{Q_{brst}, W_{s-1|s-3}\} = \alpha_s V_{s-1|s-3}$ where $\alpha_s$ are some numbers (generically, nonzero) and $G^{(k)}(\phi,\chi)$ are polynomials in derivatives of $\phi$ and $\chi$ of conformal dimension $k$ ( generically, inhomogeneous in degree and quite cumbersome) such that

$$e^{\chi-(s+1)\phi} \partial^{s-1-k} X_m G^{(k)}(\phi,\chi) \partial\chi$$

is a primary field (this is a rather stringent constraint which, nevertheless, typically has nontrivial solutions for generic $s$). For this reason, unless one of two nontriviality conditions holds, the operators $V_{s-1|s-3}$ are BRST-exact in small Hilbert space; however, if one of them are satisfied, the $W$-operators do not commute with the stress-energy tensor part of $Q_{brst}$ and therefore their overall commutators with $Q_{brst}$ no longer produce $V_{s-1|s-3}$ with the latter now being in BRST cohomology and physical. However, it is easy to see that out of 2 possible nontriviality conditions (4.70), (4.71) it is the second one that must be chosen since the first one clearly violates the $H_{-s}$-cohomology constraint. This immediately entails the gauge transformations for the $\Omega$-field:

$$\Omega_m^{a_1...a_{s-1}|b_1...b_{s-3}} \to \Omega_m^{a_1...a_{s-1}|b_1...b_{s-3}} + p_m \Lambda^{a_1...a_{s-1}|b_1...b_{s-3}} \quad (4.73)$$

that shift $V_{s-1|s-3}$ by BRST-trivial terms irrelevant for amplitudes and lead to well-known vast and powerful gauge symmetries possessed by the higher spin fields. Note that, although all the above analysis has been performed for the operators at negative cohomologies (which are simpler from the technical point of view), all the above results directly apply to the corresponding operators at isomorphic positive $H_{s-2}$-cohomologies since the explicit isomorphism between negative and positive cohomologies is BRST-invariant.

To complete our analysis of BRST on-shell constraints on the higher spin operators of $H_{s-2} \sim H_{-s}$ we shall comment on the only remaining possible source of BRST-triviality for $V_{s-1|s-3}$ coming from operators proportional to the ghost factor $\sim e^{2\chi-(s+2)\phi}$. All the hypothetical operators in the small Hilbert space with such a property are given by:

$$U_{s-1|s-3} = \Omega_m^{a_1...a_{s-1}|b_1...b_{s-3}} \oint dz c\partial\xi\partial^2\xi e^{-(s+2)\phi} R^{(2s-2)}(\phi,\chi,\sigma)$$
$$\times\psi^m\partial\psi_{b_1}...\partial^{s-3}\psi_{b_{s-3}}\partial X_{a_1}...\partial X_{s-1}e^{ipX} \tag{4.74}$$

where $R^{(2s-2)}$ is the conformal dimension $2s-2$ polynomial in derivatives of $\phi$,$\chi$ and $\sigma$ (again, homogeneous in conformal weight but not in degree). Indeed, the commutator of the matter supercurrent part of $Q_{brst}$, given by $-\frac{1}{2}\oint dw\gamma\psi_m\partial X^m$ with $U_{s-1|s-3}$ is zero since the leading order of the OPE between $\gamma\psi_m\partial X^m(w)$ and the integrand of $U_{s-1|s-3}$ at a point $z$ is nonsingular, that is, proportional to $(z-w)^0$, as is easy to check. At the same time , the commutator of $U_{s-1|s-3}$ with the ghost supercurrent part of $Q_{brst}$, given by $-\frac{1}{4}b\gamma^2$, is nonzero and is proportional to $V_{s-1|s-3}$:

$$\{Q_{brst}, U_{s-1|s-3}\} = \lambda_s V_{s-3} \tag{4.75}$$

(where $\lambda_s$ are certain numbers) provided that the coefficient $\sigma_{2s-2}$ in front of the leading OPE order of $R^{(2s-2)}$ and $b\gamma^2$ is nonzero:

$$R^{(2s-2)}(z) : b\gamma^2 : (w) \sim \frac{\sigma_{2s-2}b\gamma^2(w)}{(z-w)^{2s-2}} + O(z-w)^{2s-3}$$
$$\sigma_{2s-2} \neq 0 \tag{4.76}$$

Then, provided that the conditions

$$\lambda_s \neq 0 \tag{4.77}$$

and

$$\sigma_{2s-2} \neq 0 \tag{4.78}$$

are both satisfied, the operator $V_{s-1|s-3}$ could be trivial only if the stress-tensor part of $Q_{brst}$ commuted with $U_{s-1|s-3}$ which is only possible if (given the on-shell conditions on $\Omega$ described above)

$$G_s(z) =: c\partial\xi\partial^2\xi e^{-(s+2)\phi} R^{(2s-2)}(\phi,\chi,\sigma) : (z) \tag{4.79}$$

is a primary field. That is, the OPE of $G_s$ with the full ghost stress-energy tensor:

$$T_{gh} = \frac{1}{2}(\partial\sigma)^2 + \frac{1}{2}(\partial\chi)^2 - \frac{1}{2}(\partial\phi)^2 + \frac{3}{2}\partial^2\sigma + \frac{1}{2}\partial^2\chi - \partial^2\phi \tag{4.80}$$

is generically given by

$$T_{gh}(z)G_s(w) = \sum_{k=0}^{2s-1} \frac{y_k Y^{(-\frac{s^2}{2}-s+k)}(w)}{(z-w)^{2s+2-k}}$$
$$+\frac{(s-\frac{1}{2}s^2)G_s(w)}{(z-w)^2} + \frac{\partial G_s(w)}{(z-w)} + O(z-w)^0 \tag{4.81}$$

where $y_k$ are numbers and $Y^{(-\frac{s^2}{2}-s+k)}$ are operators of conformal dimensions $-\frac{s^2}{2} - s + k$. So the $V_{s-1|s-3}$ operators trivial only if the constraints

$$y_k = 0$$
$$k = 0, ..., 2s-1 \tag{4.82}$$

are fulfilled simultaneously with the conditions (4.77), (4.78). Clearly, for $s$ large enough the constraints (4.77), (4.78) and (4.82) are altogether too restrictive, leaving no room for any possible choice of $R^{(2s-2)}(\phi, \chi, \sigma)$, so the operators are nontrivial. To see this, note that, for any large $s$ and given $k$ the number of independent operators $Y^{(-\frac{s^2}{2}-s+k)}$ is of the order of $\sim \frac{d}{dk}(\frac{e^{a\sqrt{2s-k}}}{\sqrt{2s-k}})$ where $a$ is certain constant, since the number of conformal weight $n$ polynomials is of the order of the number of partitions of $n$ which, in turn, is given by Hardy-Ramanujan asymptotic formula for partitions, in the limit of large $n$. Summing over $k$, it is clear that the number of constraints on $G^{(2s-2)}$ is asymptotically of the order of $\frac{e^{a\sqrt{s}}}{\sqrt{s}}$ while the number of independent terms in $R^{(2s-2)}$ is of the order of $\frac{e^{a\sqrt{s}}}{s}$, so the number of constraints exceeds the number of possible operators $U_{s-1|s-3}$ by the factor of the order of $\sqrt{s}$. Therefore all the operators (4.74) with large spin values are BRST-nontrivial and physicsl, provided that the on-shell invariance constraints are satisfied. For the lower values of $s$, however, the nontriviality constraints have to be analyzed separately. For $s = 3, 4$ it can be shown that the constraints (4.77), (4.78) lead to polynomials satisfying $\lambda_s = 0$, so the appropriate higher spin operators are physical. For $5 \leq 10$ direct numerical analysis shows the incompatibility of the conditions (4.77), (4.78), (4.82) with the number of constraints exceeding the number of operators of the type (4.79) posing a potential threat of BRST-triviality, showing that operators with spins greater than 4 are physical as well.

With the on-shell BRST conditions pointed out, the next step is to analyze the scale invariance (off-shell) constraints on $V_{s-1|s-3}$. It is instructive to start with the $s = 3$ case since for $s = 3$ $\Omega_{s-1|s-3}$ is precisely the Fronsdal's field. Similarly to the closed string case, the Weyl transformation of $V_{s-1|s-3}$ is determined by the OPE coefficient in front of $\sim |z-\tau|^{-2}$ term in the operator product $lim_{z,\bar{z}\to\tau}T^{z\bar{z}}(z,\bar{z})V_{s-1|s-3}(\tau)$ where $\tau$ is on the worldsheet boundary and, as previously, the $\epsilon$-expansion setup is assumed, so $T^{z\bar{z}}\neq 0$. Starting from the transformation by $T_X=-\frac{1}{2}|\partial\vec{X}|^2$, we have

$$\begin{aligned}&\int d^2zT_X^{z\bar{z}}(z,\bar{z})\Omega_m^{a_1a_2}(p)\oint d\tau e^{-3\phi}\psi^m\partial X_{a_1}\partial X_{a_2}e^{ipX}(\tau)\\&\sim ln\Lambda\times\oint d\tau e^{-3\phi}\psi^m\partial X_{a_1}\partial X_{a_2}\\&\times e^{ipX}(\tau)[-p^2\Omega_m^{a_1a_2}(p)+2p_tp^{(a_1}\Omega_m^{a_2)t}-p^{a_1}p^{a_2}\Omega_m']\end{aligned}\tag{4.83}$$

where we introduced $\Omega'_m\equiv\eta_{a_1a_2}\Omega_m^{a_1a_2}$ (similarly, using the Fronsdal's notations, the "prime" will stand for contraction of a pair of fiber 0indices for any other higher spin field below). This gives the part of the leading order contribution to the spin 3 $\beta$-function proportional to the Fronsdal's operator in flat space. The analysis of the contributions by $T_\psi^{z\bar{z}}$ and by $T_{\beta-\gamma}^{z\bar{z}}$ is analogous to the one performed previously for the AdS graviton operator, and the result is

$$\begin{aligned}&\int d^2z(T_\psi(z,\bar{z})+T_{\beta-\gamma}(z,\bar{z}))\\&\times\Omega_m^{a_1a_2}(p)\oint d\tau e^{-3\phi}\psi^m\partial X_{a_1}\partial X_{a_2}e^{ipX}(\tau)\\&\sim-8ln\Lambda\Omega_m^{a_1a_2}\times\oint d\tau e^{-3\phi}\psi^m\partial X_{a_1}\partial X_{a_2}e^{ipX}(\tau)\end{aligned}\tag{4.84}$$

As in the case of the cosmological term appearing in the graviton's $\beta$-function, the appearance of the mass-like term in the spin 3 $\beta$-function is due to the non-cancellation of the corresponding terms in the Weyl transformation laws for the matter and for the ghost parts, which in turn is the consequence of the $H_{-3}\sim H_1$-cohomology coupling of the spin 3 operator. The term (4.84) in the $\beta$-function is *not*, however, a mass term. Namely, combined together, the contributions (4.83), (4.84) give the low-energy equations of motion for massless spin 3 field, corresponding to the special case of the Fronsdal's operator in the $AdS$ space acting on spin 3 field that is polarized along the $AdS$ boundary and is propagating parallel to the boundary. The correspondence between (4.84) and the mass-like term in the Fronsdal's operator in $AdS_{d+1}$ is exact for $d=4$; to make the correspondence precise for $d\neq 4$ requires some modification of the operators of the type of $\sim V_{s-1|s-3}$ (see the discussion below for general spin case).

The next step is to generalize this simple calculation to the general spin value and to calculate the $\beta$-functions of the frame-like fields for general $s$ values. The $V_{s-1|s-3}$-operators do not generate Fronsdal's fields for

$s \geq 4$ (but rather the derivatives of the Fronsdal's fields), and explicit expressions for $V_{s-1|t}$-operators for $0 \leq t \leq s-4$, following from cohomology constraints are generally quite complicated. For example, the manifest form of operators for $\Omega_{s-1|s-4}$-fields at canonical $(-s-1)$-picture is given by

$$\begin{aligned} &V_{s-1|s-4} \\ &= \Omega_m^{a_1...a_{s-1}|b_1...b_{s-4}}(p) \oint dz e^{-(s+1)\phi} \psi_m \partial\psi^{b_1}...\partial^{s-4}\psi^{b_{s-4}} \\ &\times \sum_{k=0}^{2s-3} T^{(2s-3-k)}(\phi)[\sum_{j=1}^{k-1} a_j \partial^j X_q \partial^{k-j} X^q + b_j \partial^{j-1}\psi_q \partial^{k-j}\psi^q] \quad (4.85) \end{aligned}$$

where $a_j$ and $b_j$ are certain coefficients and $T^{(2s-3-k)}(\phi)$ are again certain conformal dimension $2s-3-k$ inhomogeneous polynomials in the derivatives of $\phi$. The coefficients and the polynomial structures must be chosen to ensure that the integrand of (4.85) is primary field of dimension 1 and the picture-changing transformation of the operator (4.85) is nonzero, producing an operator at picture $-s$ and at cohomology $H_{-s}$, so that the hohomology condition on $V_{s-1|s-4}$ produces the zero torsion-like condition relating the frame-like fields in Vasiliev's formalism:

$$\begin{aligned} &: \Gamma V_{s-1|s-4} := V_{s-1|s-3} + \{Q_{brst}, ...\} \\ &\Omega^{s-1|s-3}(p) \sim p\Omega^{s-1|s-4}(p) \quad (4.86) \end{aligned}$$

so that the transformation of $V_{s-1|s-4}$ produces the vertex operator proportional to $V_{s-1|s-3}$ with the space-time field $\Omega^{s-1|s-3}(p) \sim p\Omega^{s-1|s-4}(p)$ given by certain first order differential operator acting on $\Omega^{s-1|s-4}(p)$. The explicit structure of this operator (giving one of the zero curvature constraints) is determined by the details of the picture-changing; for example one of the contributions to $V_{s-1|s-4}$ from the picture transformation of the $k=0$ term in $V_{s-1|s-4}$ results from the OPE contributions:

$$\begin{aligned} &e^{\phi}(z)e^{-(s+1)\phi}(w) \sim (z-w)^{s+1} e^{-s\phi}(\frac{z+w}{2}) + ... \\ &\partial X^q(z) e^{ipX}(w) \sim (z-w)^{-1} \times (-ip^q) e^{ipX}(\frac{z+w}{2}) + ... \\ &e^{\phi}(z) T^{(2s-3)}(\phi)(w) \sim (z-w)^{3-2s} e^{\phi}(\frac{z+w}{2}) + ... \quad (4.87) \end{aligned}$$

so the leading OPE order of the product of the picture-changing operator $\Gamma \sim -\frac{1}{2}e^{\phi}\psi_q \partial X^q + ...$ with $V_{s-1|s-4}$ is $\sim(z-w)^{3-s}$, so to obtain the normally ordered contribution, relevant to the picture-changing transformation of $V_{s-1|s-4}$, one has to expand the remaining field $\psi_q(z)$ of $\Gamma$ up to the order of $s-3$ around the midpoint $\frac{z+w}{2}$ which altogether produces the result proportional to $V_{s-1|s-3}$, with the space-time field proportional to the space-time derivative of $\Omega^{s-1|s-4}$ (as it is clear from the second OPE in (4.87)). There are of course many other terms in the OPE between $\Gamma$ and $V_{s-1|s-4}$ but, provided that all the coefficients and the polynomial structures in $V_{s-1|s-4}$ are chosen correctly, they all give the result proportional to $V_{s-1|s-3}$, up to BRST-exact terms and with the zero torsion condition:

$$\Omega^{s-1|s-3} \sim \partial\Omega^{s-1|s-4}$$

controlled by the picture-changing procedure. The explicit expressions for the operators with $t = s-5, s-6, \ldots$ and, ultimately, for $t = 0$ (Fronsdal's field) are increasingly complicated for general $s$. However, in order to deduce the Weyl invariance constraints on massless vertex operators for Fronsdal's fields of spin $s$, we don't actually need to know the explicit expressions for $V_{s-1|0}$. The key point here is the mutual independence of the Weyl transformations and the cohomology constraints on the vertex operators. That is, the cohomology constraints relate the Fronsdal's operator at canonical $(3-2s)$-picture and the operator for the $\Omega^{s-1|s-3}$ extra-field through

$$\begin{aligned}
&\Omega_m^{a_1\ldots a_{s-1}|b_1\ldots b_{s-3}}(p)\\
&\times \oint dz e^{-s\phi}\psi^m \partial\psi_{b_1}\ldots\partial^{s-3}\psi_{b_{s-3}}\partial X_{a_1}\ldots\partial X_{a_{s-1}}e^{ipX}\\
&= \Omega_m^{a_1\ldots a_{s-1}} : \Gamma^{s-3} : \oint dz U^{m(-2s+3)}_{a_1\ldots a_{s-1}}(p)
\end{aligned} \tag{4.88}$$

where $U$ is the indegrand of the vertex operator for the Fronsdal's field. Since $\Gamma$ is BRST and Weyl-invariant, this allows to deduce the low-energy equations of motion for the Fronsdal's fields by studying the Weyl transformations of the operators (4.74), which are much simpler. The transformations of $V_{s-1|s-3}$ by $T_X^{z\bar{z}}$ and $T_{\beta-\gamma}^{z\bar{z}}$ are computed similarly to the spin 3 case considered above. One easily finds

$$\begin{aligned}
&\int d^2z T_X^{z\bar{z}}(z,\bar{z})\Omega_m^{a_1\ldots a_{s-1}|b_1\ldots b_{s-3}}(p)\\
&\times \oint d\tau e^{-s\phi}\psi^m \partial\psi_{b_1}\ldots\partial^{s-3}\psi_{b_{s-3}}\partial X_{a_1}\ldots\partial X_{a_{s-1}}e^{ipX}
\end{aligned}$$

$$\sim ln\Lambda \oint d\tau e^{-s\phi}\psi^m \partial\psi_{b_1}...\partial^{s-3}\psi_{b_{s-3}}\partial X_{a_1}...\partial X_{a_{s-1}} e^{ipX}$$
$$\times[-p^2\Omega_m^{a_1...a_{s-1}|b_1...b_{s-3}}(p)$$
$$+p_t\Sigma_1(a_1|a_2...a_{s-1})p^{a_1}\Omega_m^{a_2...a_{s-1}t|b_1...b_{s-3}}$$
$$-\frac{1}{2}\Sigma_2(a_{s-2},a_{s-1}|a_1,...,a_{s-3})p^{a_{s-1}}p^{a_{s-2}}(\Omega'_m)^{a_1...a_{s-3}|b_1...b_{s-3}}] \tag{4.89}$$

and

$$\int d^2z T^{z\bar{z}}_{\beta-\gamma}(z,\bar{z})\Omega_m^{a_1...a_{s-1}|b_1...b_{s-3}}(p)$$
$$\times \oint d\tau e^{-s\phi}\psi^m \partial\psi_{b_1}...\partial^{s-3}\psi_{b_{s-3}}\partial X_{a_1}...\partial X_{a_{s-1}} e^{ipX}$$
$$\sim -s^2\Omega_m^{a_1...a_{s-1}|b_1...b_{s-3}}(p)$$
$$\times ln\Lambda \oint dz e^{-s\phi}\psi^m \partial\psi_{b_1}...\partial^{s-3}\psi_{b_{s-3}}\partial X_{a_1}...\partial X_{a_{s-1}} e^{ipX} \tag{4.90}$$

Here $\Sigma_1(b|a_1...a_n)$ and $\Sigma_2(b_1,b_2|a_1....a_n)$ are the Fronsdal's symmetrization operations, acting on free indices, e.g. $\Sigma_p\ (a_1,...,a_p|b_1,...,b_s)\ T^{aa_1...a_p}$ $H_a^{b_1...b_s}$ where $H$ is symmetric, symmetrizes over $a_1,...a_p;b_1,...,b_s$.

To compute the Weyl transform of the $\psi$-part, it is again helpful to use the bosonization relations for the $\psi$-fields. Since the bosonized $\varphi_i$ fields carry no background charges (as it is clear from the stress-energy tensor), the coefficient in front of the $|z-\tau|^2$ term in the OPE of $T_\psi^{z\bar{z}}(z,\bar{z})$ and $V_{s-1|s-3}(\tau)$ coincides with the conformal dimension of the $\psi$-factor: $\psi^m\partial\psi_{b_1}...\partial^{s-3}\psi_{b_{s-3}}$ which is equal to $\frac{1}{2}(s-2)^2$, so

$$\int d^2z T^{z\bar{z}}_{\psi}(z,\bar{z})\Omega_m^{a_1...a_{s-1}|b_1...b_{s-3}}(p)$$
$$\times \oint d\tau e^{-s\phi}\psi^m \partial\psi_{b_1}...\partial^{s-3}\psi_{b_{s-3}}\partial X_{a_1}...\partial X_{a_{s-1}} e^{ipX}$$
$$\sim (s-2)^2\Omega_m^{a_1...a_{s-1}|b_1...b_{s-3}}(p)$$
$$\times ln\Lambda \oint dz e^{-s\phi}\psi^m \partial\psi_{b_1}...\partial^{s-3}\psi_{b_{s-3}}\partial X_{a_1}...\partial X_{a_{s-1}} e^{ipX} + ... \tag{4.91}$$

(again, with no factor of $\frac{1}{2}$ due to the normalization chosen for the kinetic term). The last identity is true as long as the $\psi$-factor is a primary field, i.e. the appropriate on-shell conditions are imposed on $\Omega$. It is not difficult to see, however, that the contributions due to the off-shell part are generally proportional to space-time derivatives of $\Omega$ and its traces, multiplied by higher spin operators that are not of the form (4.74), so these contributions are irrelevant for $\beta$-functions of the higher spin fields of Vasiliev's type (instead, they contribute to the low-energy equations of motion of more complicated higher spin fields, such as those with mixed symmetries; so these contributions may become important in various generalizations of the Vasiliev's theory). Collecting (4.89) - (4.91) and using the cohomology constraint (4.88), we deduce that the leading order $\beta$-function for the massless Fronsdal's fields of spin $s$ is

$$\begin{aligned} &\beta_m^{a_1...a_{s-1}} \\ &= -p^2\Omega_m^{a_1...a_{s-1}}(p) + \Sigma_1(a_1|a_2,...a_{s-1})p_t p^{a_1}\Omega_m^{a_2...a_{s-1}t} \\ &-\frac{1}{2}\Sigma_2(a_{s-2},a_{s-1}|a_1,...,a_{s-3})p^{a_{s-1}}p^{a_{s-2}}(\Omega'_m)^{a_1...a_{s-3}} \\ &-4(s-1)\Omega_m^{a_1...a_{s-1}} \end{aligned} \quad (4.92)$$

The appearance of the mass-like terms is related to the emergence of the curved geometry already observed in the spin 2 case. Namely, vanishing of the $\beta$-function (4.92) gives, in the leading order, the low-energy effective equations of motion on $\Omega$ given by

$$\hat{F}_{AdS}\Omega = 0 \quad (4.93)$$

where $\hat{F}_{AdS}$ is the Fronsdal's operator in $AdS_{d+1}$ space, (exactly for $d=4$ and with some modifications in other dimensions) which action is restricted on higher spin fields $\Omega$ polarized along the $AdS$ boundary. Indeed, the explicit expression for the Fronsdal's operator in $AdS_{d+1}$, acting on symmetric spin $s$ fields polarized along the boundary is:

$$\begin{aligned} &(\hat{F}_{AdS}\Omega)^{a_1...a_s} = \nabla_A\nabla^A\Omega^{a_1...a_s} - \Sigma_1(a_1|a_2...a_s)\nabla_t\nabla^{(a_1}\Omega^{a_2...a_s t)} \\ &+\frac{1}{2}\Sigma_2(a_1,a_2|a_3,...,a_s)\nabla^{a_1}\nabla^{a_2}(\Omega')^{a_3...a_s} \\ &-m_\Omega^2\Omega^{a_1...a_s} + 2\Sigma_2\Lambda g^{a_1a_2}(\Omega')^{a_3...a_s} \\ &m_\Omega^2 = -\Lambda(s-1)(s+d-3) \end{aligned} \quad (4.94)$$

where $A=(a,\alpha)$ is the $AdS_{d+1}$ space-time index (with the latin indices being along the boundary and $\alpha$ being the radial direction).

The cosmological constant in our units is fixed $\Lambda = -4$, to make it consistent with the Weyl transform of the $AdS$ graviton operator. In what follows we shall ignore the last term in this operator since, in the string theory context, it is related to the higher-order (cubic) contributions to the $\beta$-function, which are beyond the leading order Weyl invariance constraints.

For the remaining part, consider the box ($\nabla^2$) of $\Omega$ first. It is convenient to use the Poincare coordinates for $AdS$:

$$ds^2 = \frac{R^2}{y^2}(dy^2 + dx_a dx^a) \tag{4.95}$$

With the Christoffel's symbols:

$$\Gamma^y_{a_1 a_2} = -\Gamma^y_{yy}\delta_{a_1 a_2} = -\frac{1}{y}\delta_{a_1 a_2} \tag{4.96}$$

one easily computes:

$$\begin{aligned}
&\nabla_A \nabla^A \Omega_{a_1 \dots a_s}(x) \equiv (\nabla_a \nabla^a + \nabla_y \nabla^y)\Omega_{a_1 \dots a_s}(x) \\
&= (\partial_a \partial^a - \Lambda s(s+d))\Omega_{a_1 \dots a_s}
\end{aligned} \tag{4.97}$$

Substituting into the AdS Fronsdal's operator in the momentum space (with the Fourier transformed boundary coordinates) gives

$$\begin{aligned}
&(\hat{F}_{AdS}\Omega(p))^{a_1 \dots a_s} \\
&= -p^2 \Omega_m^{a_1 \dots a_{s-1}}(p) + \Sigma_1(a_1|a_2 \dots a_{s-1}) p_t p^{a_1} \Omega_m^{a_2 \dots a_{s-1} t} \\
&-\frac{1}{2}\Sigma_2(a_{s-2}, a_{s-1}|a_1, \dots a_{s-3}) p^{a_{s-1}} p^{a_{s-2}} (\Omega'_m)^{a_1 \dots a_{s-3}} \\
&+\Lambda(s+3-d)\Omega_m^{a_1 \dots a_{s-1}}
\end{aligned} \tag{4.98}$$

Thus the $\beta$-functions for the $V_{s-1|0}$ vertex operators coincide with $AdS$ Fronsdal operators precisely for $AdS_5$ case ($d = 4$). For other values of $d$ the string theoretic calculation of the mass-like factor $m_\Omega^2 \sim \Lambda(s-1)$ is still proportional to $s$, but there is a discrepancy proportional to $d - 4$. This discrepancy can always be cured, however, by suitable modification of the $\psi$-part of the vertex operators of the type (4.74). This modification typically involves the shift of the canonical picture of the operator for the Fronsdal's field from $2s - 3$ to $2s - 3 + |d - 4|$ and is somewhat tedious, but straightforward, with the explicit form depending on $d$. However, the shift doesn't change the order of the cohomology, which is still $H_{s-2} \sim H_{-s}$ for each value of $s$. The Regge-style behavior (4.98) of the mass-like terms in Fronsdal operators is thus the consequence of the cohomology structure of the higher spin vertices in the "larger" string theory. In the following section, we shall use the frame-like operators, constructed above, to describe massless higher-spin interactions in the AdS space (unless stated otherwise, we will limit ourselves to the case of $s = 3$ with the space-time fields are polarized along the $AdS$ boundary).

## 4.5 3-Point Amplitude and 3-Derivative Vertex

In this section we use the vertex operator formalism, developed in the previous section, to compute the cubic coupling of massless spin 3 fields. Below we shall limit ourselves to the 3-derivative contributions corresponding to the Berends, Burgers and Van Dam (BBD) [17, 85] type vertex in the field theory limit. The first step is to choose the ghost pictures of the operators to ensure the correct ghost number balance, i.e. so that the correlator has total $\phi$-ghost number $-2$, $b-c$ ghost number $+3$ and $\chi$-ghost number $+1$. This requires two out of three operators to be taken unintegrated at negative pictures and the third one at positive picture (note that higher spin operators at positive pictures are always integrated). It is convenient to take unintegrated operators at the minimal ghost picture $-3$, i.e. we shall use the $V^{(-3)}$ operator for $\omega^{2|0}$. Then the remaining integrated operator must be taken at picture $+5$, and only the terms proportional to the ghost factor $\sim ce^{\chi+4\phi}$ will contribute, while the terms proportional to $\sim \partial cce^{2\chi+3\phi}$ and to $\sim e^{5\phi}$ will drop out as they don't satisfy the balance of ghosts. It is therefore appropriate to choose the operator for $\omega^{2|2}$ for the third operator (which minimal positive picture is $+3$) and to apply the picture changing transformation twice to bring it to the picture $+5$. The result is given by

$$\{Q,\xi\{Q,\xi K\circ\omega_m^{ab|cd}(p)\oint de^{3\phi}F_{abcd}^{m(\frac{17}{2})}\}\}=V_1(p)+V_2(p)$$
$$\equiv\omega_m^{ab|cd}(p)\oint du(u_0-u)^8ce^{\chi+4\phi+ipX}R^m_{ab|cd}(u) \tag{4.99}$$

where

$$V_1(p)=\frac{3}{64}\omega_m^{ab|cd}(p)\oint du(u_0-u)^8ce^{\chi+4\phi+ipX}L_{abcd}^{m(9)}$$
$$\times\{\frac{24}{11!}P^{(11)}_{2\phi-2\chi-\sigma}(\frac{1}{8}P^{(2)}_{\chi}+\frac{1}{8}P^{(2)}_{2\phi-2\chi-\sigma}-\frac{1}{4}P^{(1)}_{\chi}P^{(1)}_{2\phi-2\chi-\sigma}$$
$$-12P^{(11)}_{\phi-\chi}P^{(1)}_{2\phi-2\chi-\sigma}-12(P^{(1)}_{\phi-\chi})^2)$$
$$+\frac{1}{11!}P^{(12)}_{2\phi-2\chi-\sigma}P^{(1)}_{-\frac{3}{2}\phi+\frac{55}{4}\chi+\frac{11}{4}\sigma}-\frac{102}{13!}P^{(13)}_{2\phi-2\chi-\sigma}\} \tag{4.100}$$

and

$$V_2(p) = \frac{1}{9!-8!}\sum_{n=0}^{7} 2^{n-7} \sum_{\{l,m,p,q\geq 0; l+m+p+q=8-n\}} \sum_{r=0}^{p}\sum_{a=0}^{l}$$
$$\sum_{b=0}^{q}\sum_{N=0}^{a+b+r+5}\{\frac{(-1)^{a+b+p+q}N!}{m!l!(p-r)!r!(N-r)!(5+a+b+r-N)!}$$
$$\times\omega_m^{ab|cd}(p)\oint du(u_0-u)^8 ce^{\chi+4\phi}\partial^{(p-r)}L_{abcd}^{m(N+9)}$$
$$\times\partial^{(m)}P^{n|8}_{2\phi-2\chi-\sigma|\chi}B^{l-a|l}_{\chi|\phi-\chi}B^{q-b|q}_{3\phi+\chi|\phi-\chi}B^{(5+a+b+r-N)}_{\phi-\chi}(u)\} \quad (4.101)$$

Here $u_0$ is an arbitrary point on the boundary and

$$F_{abcd}^{m(\frac{17}{2})} = (\psi^m\partial^2\psi_c\partial^3\psi_d\partial X^a\partial X_b - 2\psi^m\partial\psi_c\partial^3\psi_d\partial X_a\partial^2 X_b$$
$$+\frac{5}{8}\psi^m\partial\psi_c\partial^2\psi_d\partial X_a\partial^3 X_b + \frac{57}{16}\psi^m\partial\psi_c\partial^2\psi_d\partial^2 X_a\partial^2 X_b)e^{ipX} \quad (4.102)$$

is dimension $\frac{17}{2}$ primary field (given the on-shell conditions on $\omega$); conformal dimension $N+9$ fields are defined as the OPE terms in the product of $F_{abcd}^{m(\frac{17}{2})}$ with the matter supercurrent $G = -\frac{1}{2}\psi_n\partial X^n$ on the worldsheet:

$$G_m(z)F_{abcd}^{m(\frac{17}{2})}(w) = \sum_{N=0}^{\infty}(z-w)^{N-1}L_{abcd}^{m(N+9)}(w) \quad (4.103)$$

or manifestly

$$L_{abcd}^{m(N+9)}$$
$$= \frac{e^{ipX}}{N!}\{\frac{15}{8}(\partial X_a\partial^3 X_b + \frac{171}{16}\partial^2 X_a\partial^2 X_b)(\partial^{(N+1)}X^m\partial\psi_c\partial^2\psi_d$$
$$-\frac{1}{N+1}\partial^{(N+2)}X_c\psi^m\partial^2\psi_d$$
$$+\frac{2}{(N+1)(N+2)}\partial^{(N+3)}X_d\psi^m\partial\psi_c)$$
$$+\psi^m\partial\psi_c\partial^2\psi_d(-\frac{15}{8(N+1)}\partial^{N+1}\psi_a\partial^3 X_b$$
$$-\frac{45}{4(N+1)(N+2)(N+3)}\partial^{N+3}\psi_b\partial X_a$$
$$-\frac{171}{8(N+1)(N+2)}\partial^{N+2}\psi_a\partial^2 X_b)$$

$$
\begin{aligned}
&+3\partial X_a \partial X_b (\partial^{(N+1)} X^m \partial^2 \psi_c \partial^3 \psi_d \\
&-\frac{2}{(N+1)(N+2)}\partial^{(N+3)} X_c \psi^m \partial^3 \psi_d \\
&-\frac{6}{(N+1)(N+2)(N+3)}\partial^{(N+4)} X_d \partial^2 \psi_c \psi^m \\
&-\frac{3}{N+1}\psi^m \partial^2 \psi_c \partial^3 \psi_d \partial^{(N+1)} \psi_{(a} \partial X_{b)}) \\
&-6\partial X_a \partial^2 X_b (\partial^{(N+1)} X^m \partial \psi_c \partial^3 \psi_d \\
&-\frac{1}{N+1}\partial^{(N+2)} X_c \psi^m \partial^3 \psi_d \\
&-\frac{6}{(N+1)(N+2)(N+3)}\partial^{(N+4)} X_d \partial \psi_c \psi^m) \\
&+6\psi^m \partial \psi_c \partial^3 X_d \\
&\times (\frac{1}{N+1}\partial^{(N+1)} \psi_a \partial^2 X_b + \frac{2}{(N+1)(N+2)}\partial^{(N+2)} \psi_b \partial X_a) \\
&+N : \partial^{N-1} G_m F^{m(\frac{17}{2})}_{abcd} : (1-\delta_{0;N}) - i : (p^n \partial^N \psi_n) F^{m(\frac{17}{2})}_{abcd} :\}
\end{aligned}
\tag{4.104}
$$

We are now prepared to analyze the 3-point function given by

$$
\begin{aligned}
&A(p,k,q) = \omega_n^{s_1 s_2}(p)\omega_p^{t_1 t_2}(k)\omega_m^{ab|cd}(q) \oint du (u-u_0)^8 \\
&< c\partial X_{s_1} \partial X_{s_2} \psi^n e^{ipX}(z) c\partial X_{t_1} \partial X_{t_2} \psi^n e^{ikX}(w) \\
&c e^{\chi + 4\phi + iqX} R^m_{ab|cd}(u) >
\end{aligned}
\tag{4.105}
$$

Using the $SL(2,R)$ symmetry, it is convenient to set $z \to \infty, u_0 = w = 0$. For the notation purposes, however, it is convenient to retain $z$ and $w$ in our notations for a time being. We start with computing the "static" exponential ghost part of the correlator. Simple calculation gives

$$
\begin{aligned}
&< ce^{-3\phi}(z) ce^{-3\phi}(w) ce^{4\phi + \chi(u)} > \\
&= (z-w)^{-8}(z-u)^{13}(w-u)^{13} \to z^5 (w-u)^{13}
\end{aligned}
\tag{4.106}
$$

where we substituted the $z \to \infty$ limit. Next, consider the $\psi$-part of the correlator. The expression for $R^m_{ab|cd}$ contains two types of terms: those that are quadratic in $\psi$ and those that are quartic $\psi$ . Since the remaining two spin 3 operators are linear in $\psi$, only the quadratic terms contribute to the correlator. Note that all the terms quadratic in $\psi$ are also cubic in

$\partial X$. So the pattern for the $\psi$-correlators is

$$\begin{aligned}
&< \psi^n(z)\psi^p(w) : \partial^{(P_1)}\psi_c\partial^{(P_2)}\psi_d : (u) > \\
&= P_1!P_2!(\frac{\eta_d^n\eta_c^p}{z^{P_2+1}(w-u)^{P_1+1}} - \frac{\eta_c^n\eta_d^p}{z^{P_1+1}(w-u)^{P_2+1}})
\end{aligned} \tag{4.107}$$

where, according to the manifest expression for $R^m_{ab|cd}$ the numbers $P_1$ and $P_2$ can vary from 0 to $N+3$ (and $N_{max} = 8$). Next, consider the $X$-part. As we limit ourselves to just three-derivative terms, it is sufficient to compute the terms linear in momentum (since the $\omega^{2|2}$ field already contains 2 derivatives out of 3). The $X$-factor is a combination of the 3-point correlators of the type $\sim(\omega^{2|0})^2\omega^{2|2}$ $< (\partial X)^2 e^{ipX}(z)(\partial X)^2 e^{ikX}(w)\partial^{(M_1)}X\partial^{(M_2)}X\partial^{(M_3)}Xe^{iqX} >$ with different values of $M_1$, $M_2$ and $M_3$. Straightforward computation gives:

$$\begin{aligned}
&lim_{z\to\infty}\omega_n^{s_1s_2}(p)\omega_p^{t_1t_2}(k)\omega_m^{ab|cd}(q) \\
&\times < \partial X_{s_1}\partial X_{s_2}e^{ipX}(z)\partial X_{t_1}\partial X_{t_2}e^{ikX}(w) \\
&\partial^{(M_1)}X_a\partial^{(M_2)}X_b\partial^{(M_1)}X^m e^{iqX}(u) > \\
&= M_1!M_2!M_3!\omega_n^{s_1s_2}(p)\omega_p^{t_1t_2}(k)\omega_m^{ab|cd}(q) \\
&\times\{\frac{2iq_{t_2}\eta_{s_1a}\eta_{s_2b}\eta_{t_1}^m}{z^{2+M_1+M_2}(w-u)^{2+M_3}} \\
&+iq_{t_2}\eta_{s_1a}\eta_{s_2}^m\eta_{t_1b} \\
&(\frac{1}{z^{2+M_1+M_3}(w-u)^{2+M_2}} + \frac{1}{z^{2+M_2+M_3}(w-u)^{2+M_1}}) \\
&-\frac{2ik_{s_2}\eta_{t_1a}\eta_{t_2b}\eta_{s_1}^m}{z^{3+M_3}(w-u)^{1+M_1+M_2}} \\
&-ik_{s_2}\eta_{t_1a}\eta_{t_2}^m\eta_{s_1b} \\
&(\frac{1}{z^{3+M_1}(w-u)^{1+M_2+M_3}} + \frac{1}{z^{3+M_2}(w-u)^{1+M_1+M_3}}) \\
&-\frac{1}{M_3}\eta_{s_1t_1}\eta_{s_2a}\eta_{t_2b} \\
&\times(ik^m(\frac{1}{z^{3+M_1}(w-u)^{1+M_2+M_3}} + \frac{1}{z^{3+M_2}(w-u)^{1+M_1+M_3}}) \\
&+ip^m(\frac{1}{z^{3+M_1+M_3}(w-u)^{1+M_2}} \\
&+\frac{1}{z^{3+M_2+M_3}(w-u)^{1+M_1}}))
\end{aligned}$$

$$+ip_b\eta_{s_1t_1}\eta_{s_2a}\eta^m_{t_2}(\frac{1}{M_2}\frac{1}{z^{3+M_1}(w-u)^{1+M_2+M_3}}$$
$$+\frac{1}{M_1}\frac{1}{z^{3+M_2}(w-u)^{1+M_1+M_3}})$$
$$+\frac{ip_b\eta_{s_1t_1}\eta^m_{s_2}\eta_{t_2a}}{z^{3+M_3}(w-u)^{1+M_1+M_2}}(\frac{1}{M_1}+\frac{1}{M_2}) \qquad (4.108)$$

Comparing this with the explicit expression for the $\omega^{2|2}$ vertex operator, it is easy to notice that, while the static ghost factor is proportional to $\sim z^5$, the $z$-asymptotics of the $\psi$-correlator is $\sim \frac{1}{z}+O(\frac{1}{z^2})$ and the asymptotics for the $X$-correlator is $\sim \frac{1}{z^4}+O(\frac{1}{z^5})$. This means that only the terms proportional to $\sim z^0$ contribute to the interaction vertex. Terms proportional to negative powers of $z$ disappear in the limit $z\to\infty$ and correspond to pure gauge contributions. There are no terms proportional to positive powers of $z$ (their presence would be a signal of problems with the gauge invariance). Moreover the $z$-asymptotics further simplifies the analysis of the ghost polynomials in the expressions for the $\omega^{2|2}$ operator; namely, all the polynomials have to couple to the $ce^{-3\phi}$ ghost exponent of the $\omega^{2|0}$ operator sitting at $w$ as any couplings of these polynomials with the operator sitting at $z$ produce contributions vanishing in the $z\to\infty$ limit.

Combining (4.103), (4.104), (4.107) and (4.108) we arrive to the following expression for the main matter building block for the matter part of the correlator:

$$lim_{z\to\infty}\omega_n^{s_1s_2}(p)\omega_p^{t_1t_2}(k)\omega_m^{ab|cd}(q)$$
$$\times <\psi^n\partial X_{s_1}\partial X_{s_2}e^{ipX}(z)\psi^p\partial X_{t_1}\partial X_{t_2}e^{ikX}(w)L^{m(N+9)}_{abcd}(u)>$$
$$=z^{-5}(w-u)^{-9-N}A_N(p,k,q)$$
$$A_N(k,p,q)=\omega_n^{s_1s_2}(p)\omega_p^{t_1t_2}(k)\omega_m^{ab|cd}(q)$$
$$\times\{\eta^{nm}\eta_{pd}(-72(N+5)\eta^{s_1a}\eta^{s_2b}\eta^{t_1c}q^{t_2}$$
$$+(72(N+5)-\frac{45}{4}(N+1)^2(N+2)-144)\eta^{t_1a}\eta^{s_1b}\eta^{t_2c}k^{s_2}$$
$$+(144-\frac{45}{4}N(N+1))\eta^{s_1t_1}\eta^{s_2a}\eta^{t_2b}k^c$$
$$+(\frac{45}{4}N(N+1)^2-72(N+4))\eta^{s_1t_1}\eta^{s_2a}\eta^{t_2c}p^b)$$
$$+Symm(m,a,b)\} \qquad (4.109)$$

Using the manifest expression for the $V_{2|2}$ vertex operator in terms of $L_{9+N}$ and their derivatives, it is now straightforward to calculate the cubic coupling. First of all, it is immediately clear that only the $V_2$-part of $V_{2|2}$

contributes to the overall correlator. No terms from $V_1$ contribute since, as it was pointed out above, all the ghost polynomials entering $V_{2|2}$ must be completely absorbed by the ghost exponent $\sim ce^{-3\phi}$ located at $w$ (no couplings to the exponent at $z$ are allowed as they would result in contributions vanishing at $z \to \infty$). At the same time all the terms in $V_1$ carry the factors of $P^{(n)}_{2\phi-2\chi-\sigma}$ ($n = 11, 12, 13$) which cannot be absorbed by $ce^{-3\phi}$ (i.e. their OPE's with $ce^{-3\phi}$) are less singular than $(z-w)^{-n}$). Indeed, since $e^{2\phi-2\chi}b(z)ce^{-3\phi}(w) \sim (z-w)^5 e^{-\phi-2\chi}(w) + O(z-w)^6$, clearly for $n \geq 5$

$$\partial^{(n)}(e^{2\phi-2\chi}b)(z) \equiv ce^{-3\phi}(w) \equiv :B^{(n)}_{2\phi-2\chi-\sigma}e^{2\phi-2\chi}b : ce^{-3\phi}(w)$$
$$\sim \frac{n!}{(n-5)!} : B^{(n-5)}_{2\phi-2\chi-\sigma}e^{-\phi-2\chi} : (w) + O(z-w) \tag{4.110}$$

implying that

$$B^{(n)}_{2\phi-2\chi-\sigma}(z)ce^{-3\phi}(w) \sim O(\frac{1}{z^5}) \tag{4.111}$$

i.e. no complete contractions for $n \geq 6$. Next, combining with the expression for $V_2(q)$ we obtain the following result for the overall correlator:

$$\frac{6}{9!-8!}\sum_{n=0}^{7} 2^{n-7} \sum_{\{l,m,p,q\geq 0; l+m+p+q=8-n\}} \sum_{r=0}^{p}\sum_{b=0}^{q}\sum_{N=l+b+r+2}^{l+b+r+5}$$
$$\{\frac{(-1)^{1+b+m+q+r+N}N!(N+8+p-r)!\prod_{j=0}^{m-1}(n+j)!}{m!l!(p-r)!r!(N-r)!(5+l+b+r-N)!}$$
$$\times((N-l-b-r-2)!(N+8)!)^{-1}$$
$$\times\alpha^{-3;0;1}_{3;1;0|1;-1;0}(q-b|q)\alpha^{-3;0;1}_{2;-2;-1|0;1;0}(n|8)A_N(p,k,q)\} \tag{4.112}$$

where we used the fact that the only nonzero contributions from the summation over $a$ are the terms with $a = l$ for which $P^{l-a|l}_{\chi|\phi-\chi} = P^{0|l}_{\chi|\phi-\chi} = (-1)^l l!$; while for $a \neq l$ $P^{l-a|l}_{\chi|\phi-\chi}$ are the polynomials in $\chi$ of dimension $l-a$ which aren't contractible with $ce^{-3\phi}$. The numbers $\alpha^{A_3;B_3;C_3}_{A_1;B_1;C_1|A_2;B_2;C_2}(n|N)$ appearing in (4.112) are the coefficients in front of the leading order terms in the operator products

$$P^{n|N}_{A_1\phi+B_1\chi+C_1\sigma|A_2\phi+B_2\chi+C_2\sigma}(z)e^{A_3\phi+B_3\chi+C_3\sigma}(w)$$
$$\sim \frac{\alpha^{A_3;B_3;C_3}_{A_1;B_1;C_1|A_2;B_2;C_2}(n|N)}{(z-w)^n}e^{A_3\phi+B_3\chi+C_3\sigma}(w) \tag{4.113}$$

Finally, substituting for $A_N(p,k,q)$ and evaluating the series we obtain the following answer for the cubic coupling:

$$A(p,k,q) = \frac{691072283467i}{720}\omega_n^{s_1s_2}(p)\omega_p^{t_1t_2}(k)\omega_m^{ab|cd}(q)$$
$$\times\{\eta^{nm}\eta_{pd}(\frac{1}{36}\eta^{s_1a}\eta^{s_2b}\eta^{t_1c}q^{t_2} + \frac{4}{3}\eta^{t_1a}\eta^{s_1b}\eta^{t_2c}k^{s_2}$$
$$+\frac{1}{12}\eta^{s_1t_1}\eta^{s_2a}\eta^{t_2b}k^c - \eta^{s_1t_1}\eta^{s_2a}\eta^{t_2c}p^b)$$
$$+Symm(m,a,b)\} \qquad (4.114)$$

This concludes the calculation of the 3-derivative part of the cubic vertex from string theory. Inclusion of the appropriate Chan-Paton's indices is straightforward and leads to vertices of the type considered in [17,85].

## 4.6 Quartic Interactions and Nonlocalities

Quartic interactions in higher-spin theories are holographically dual to conformal blocks on the AdS boundary. These interactions are known to be genuinely nonlocal and these nonlocalities, in general, cannot be removed by field redefinitions. Our goal in this section is to understand the origin of these nonlocalities from string theory point of view. In string theory, the structure of the higher-spin quartic interaction is related to the four-point worldsheet correlators. Typically, the four point correlators of "standard" vertex operators (elements of $H_0$) lead to Veneziano amplitudes, bearing no trace of nonlocality and leading to perfectly local quartic terms in the low-energy effective action. In case of the amplitudes involving the operatorsof $H_n \sim H_{n-2}$ for higher-spin fields, the situation changes radically because of the different $b-c$ ghost content of these operators. That is, consider the standard Veneziano amplitude. Its well-known structure results from the four-point function involving one integrated and three unintegrated vertices, that ensure the cancellation of the $b-c$ ghost number anomaly due to the background charge of the $b-c$ system. Three $c$-ghost insertions on the sphere lead to the standard $SL(2,R)$ volume factor for open strings and $SL(2,C)$ for closed strings. The single worldsheet integration then leads to the amplitude structure $\sim \frac{\Gamma\Gamma}{\Gamma}$ for open strings and $\sim \frac{\Gamma\Gamma\Gamma}{\Gamma\Gamma\Gamma}$ for closed strings, where $\Gamma$ are the gamma-functions in Mandelstam variables. The simple poles in the amplitudes occur at non-positive integer values of the Mandelstam variables; in particular, residues at massless poles determine the quartic interactions of space-time field in the low-energy effective action (in the leading order of $\alpha'$). Things change significantly with the higher-spin operators entering the game. Typically, a four-point amplitude involving the spin $n+2$ operators of $H_n \sim H_{-n-2}(n>0)$ must contain at least two vertex operators at positive picture representation, in order to cancel the background charge of the $\beta-\gamma$ system, equal to 2. As we have seen above, such operators do not admit a representation at unintegrated $b-c$ picture, so the 4-point amplitudes of the higher-spin operators involve at least 2 worldsheet integrations. Because of that, the resulting expressions for the amplitudes develop the "anomalous" factors (in addition to the standard Veneziano structure) leading to the appearance of the nonlocalities in the quartic interactions of the higher-spin fields. Below we shall consider a few examples of how this scenario for the higher-spin nonlocalities unfolds in practice. Consider the 4-point amplitude, describing the interaction of massless higher-spin fields with the spin values $s_1$,$s_2$, $s_3$ and $s_4$. The structure of this amplitude becomes relatively simple if the spin values are subject to the constraint:

$$s_1+s_2=s_3+s_3+2 \tag{4.115}$$

If this constraint is satisfied, the operators (4.74), with the linearized on-shell constraints on the space-time fields describing the propagation of massless frame-like higher-spin modes in AdS (along the AdS boundary), can be taken at their canonical pictures. Otherwise, the calculation of the amplitude would require the insertions of the picture-changing operators, related to the generalized zero-torsion constraints for frame-like fields, making the whole computation quite messy. Nevertheless, the special case (4.115) is already general enough to grasp the architectural basics of the higher-spin quartic interactions in the AdS space. With the constraints (4.115) satisfied, it is natural to choose the spins $s_1$ and $s_2$ operators to be in positive cohomologies and those of spins $s_3$ and $s_4$ $b-c$ local and in the negative cohomologies, so that the open string amplitude for the quartic has the form:

$$\begin{aligned}
&A(s_1...s_4|p_1...p_4)\\
&=< K\circ\int_0^1 d\xi_1 e^{(s_1-2)\phi}\prod_{i,j=1}^{s_1-2}\partial^{i-1}\psi_{\mu_i}\partial X_{\alpha_j}e^{ip_1X}(\xi_1)\\
&K\circ\int_0^{\xi_1} d\xi_2 e^{(s_2-2)\phi}\prod_{i,j=1}^{s_2-2}\partial^{i-1}\psi_{\nu_i}\partial X_{\beta_j}e^{ip_2X}(\xi_2)
\end{aligned}$$

$$ce^{-s_3\phi}\prod_{i,j=1}^{s_3-2}\partial^{i-1}\psi_{\rho_i}\partial X_{\gamma_j}e^{ip_3X}(\xi_3=0)$$

$$ce^{-s_4\phi}\prod_{i,j=1}^{s_4-2}\partial^{i-1}\psi_{\sigma_i}\partial X_{\delta_j}e^{ip_4X}(w\to\infty)> \tag{4.116}$$

where we have partially fixed the $SL(2,R)$ symmetry, choosing the negative cohomology operators at 0 and $\infty$. Now it is clear that, according to the ghost number selection rules, up to the interchange of $\xi_1 \leftrightarrow \xi_2$ the $s_1$ and $s_2$ operators each of two $K$-transformation's only contributions to the correlator are:

1) the one proportional to $\sim e^{\chi+(s_1-3)\phi}$ ghost factor in $K\circ V_{s_1}$.

2)the one proportional to $\sim e^{(s_1-2)\phi}$ ghost factor in $K\circ V_{s_2}$.

The straightforward evaluation of these factors in the $K$-transformations, using the symmetry in $\alpha$ and $\mu$ indices, gives:

$$\begin{aligned}
&K\circ V_s(p)=V_s^{(1)}(p)+V_s^{(2)}(p)V_s^{(1)}(\zeta)\\
&=-\frac{1}{2}\Omega^{\alpha(s-1)|\mu_{(s-3)}}(p)\int dz(\zeta-z)^{2s-4}:e^{ipX}\\
&\times\{\sum_{q=1}^{s-2}(-1)^q(q-1)![T^\perp\psi^{\mu_q}_{\mu(s-2)}]\\
&\times([T^{||}X]^{\alpha_{s-1}}_{\alpha(s-1)}\eta^{\mu_q\alpha_{s-1}}(1-s)B^{(s+q+2)}_{\phi-\chi}\\
&+[T^{||}X]_{s-1}(-ip)^{\mu_q}B^{(s+q+1)}_{\phi-\chi})\\
&+\sum_{q=1}^{s-2}\sum_{r=0}^{s+q}(-1)^q(q-1)![T^\perp\psi^{\mu_q}_{\mu(s-2)}]\\
&\times([T^{||}X]_{s-1}B^{(s+q-r)}_{\phi-\chi}\frac{\partial^{r+1}X_{\mu_q}}{r!})\\
&-\sum_{r=0}^{s+2}(-1)^q(q-1)![T^\perp\psi_{\mu(s-2)}]\\
&\times([T^{||}X]^{\alpha_{s-1}}_{\alpha(s-1)}]B^{(s+2-r)}_{\phi-\chi}\frac{\partial^r\psi_{\alpha_{s-1}}}{r!})\\
&-\sum_{r=0}^{s+2}(-1)^q(q-1)![T^\perp\psi_{\mu(s-2)}]\\
&\times([T^{||}X]_{\alpha(s-1)}]B^{(s+1-r)}_{\phi-\chi}\frac{p^\alpha\partial^r\psi_\alpha}{r!}\}:(z))
\end{aligned}\tag{4.117}$$

$$V_s^{(2)}(\zeta) = -\frac{1}{4}\Omega^{\alpha(s-1)|\mu(s-3)}(p)$$
$$\times\int dz(\zeta - z)^{2s-4} : e^{ipX} B^{(2s-4)}_{2\phi-2\chi-\sigma}[T^{\perp}\psi_{\mu(s-2)}]$$
$$\times[T^{||}X]_{\alpha(s-1)} : (z)] \quad (4.118)$$

where we have adopted the following notations:

$$[T^{||}X]_{\alpha(s)} =: \prod_{i=1}^{s}\partial X_{\alpha_i} :$$
$$[T^{||}X]^{\alpha_q}_{\alpha(s)} =: \prod_{i=1;i\neq q}^{s}\partial X_{\alpha_i} :$$
$$[T^{\perp}\psi_{\mu(s)}] =: \prod_{j=1}^{s}\partial^{j-1}\psi_{\mu_j} :$$
$$[T^{\perp}\psi^{\mu_q}_{\mu(s)}] =: \prod_{j=1;j\neq q}^{s}\partial^{j-1}\psi_{\mu_j} :$$
$$\Omega^{\alpha(s)|\mu(t)}(p) = \Omega^{\alpha_1\ldots\alpha_s|\mu_1\ldots\mu_t}(p) \quad (4.119)$$

Given the ghost number selection rules, the terms contributing to the 4-point correlator are given by:

$$< V^{(1)}_{s_1}(\zeta_1)V^{(2)}_{s_2}(\zeta_2)V_{s_3}(0)V_{s_2}(w\to\infty) > +\xi_1\to\xi_2; s_1\to s_2$$

The overall correlator stems from the contributions of $\psi$-dependent, $X$-dependent and ghost-dependent ingredients of the higher-spin vertex operators. We start from the $\psi$-part. It is given by the two patterns; the first one is

$$< T^{\perp}\psi^{\mu_q}_{\mu(s_1-2)}(\zeta_1)T^{\perp}\psi_{\nu(s_2-2)}(\zeta_2)T^{\perp}\psi_{\rho(s_3-2)}(0)$$
$$T^{\perp}\psi_{\sigma(s_4-2)}(w\to\infty) >= \sum_{s_1-3,s_2-2,s_3-2,s_4-2|\{\lambda_{ij}\}}$$
$$\sum_{\frac{(s_1-2)(s_1-3)}{2}-q;\frac{(s_2-2)(s_2-3)}{2};\frac{(s_3-2)(s_2-3)}{2};\frac{(s_4-2)(s_4-3)}{2}|\{|\Delta_{ij}\}}$$
$$\Xi^{(1-\psi)}_{q;\mu_1\ldots\mu_{s_1-2}\nu_1\ldots\nu_{s_2-2}\rho_1\ldots\rho_{s_3-2}\sigma_1\ldots\sigma_{s_4-2}}(\{\lambda\},\{\Delta\})$$
$$\times(\zeta_1-\zeta_2)^{-(\Delta_{12}+\delta_{21}+\lambda_{12})}\zeta_1^{-(\Delta_{13}+\Delta_{31}+\lambda_{13})}$$
$$\times\zeta_2^{-(\Delta_{23}+\Delta_{32}+\lambda_{23})}w^{\frac{-(s_4-2)(s_4-1)}{2}-\Delta_{41}-\Delta_{42}-\Delta_{43}}$$

$$\equiv \sum_{s_1-3,s_2-2,s_3-2,s_4-2|\{\lambda_{ij}\}} \sum^{(s_1-3;s_2-2;s_3-2;s_4-2)}_{ord.\{i,j,k\},\{l,m,n\},\{p,q,r\},\{s,t,u\}}$$
$$\sum_{\frac{(s_1-2)(s_1-3)}{2}-q;\frac{(s_2-2)(s_2-3)}{2};\frac{(s_3-2)(s_2-3)}{2};\frac{(s_4-2)(s_4-3)}{2}|\{|\Delta_{ij}\}}$$
$$(\zeta_1-\zeta_2)^{\Delta_{12}+\delta_{21}+\lambda_{12}}\zeta_1^{\Delta_{13}+\Delta_{31}+\lambda_{13}}$$
$$\times\zeta_2^{\Delta_{23}+\Delta_{32}+\lambda_{23}}w^{\frac{(s_4-2)(s_4-1)}{2}+\Delta_{41}+\Delta_{42}+\Delta_{43}}$$
$$\times\eta_{\mu_{i_1}\nu_{l_1}}...\eta_{\mu_{i_{\lambda_{12}}}\nu_{l_{\lambda_{12}}}}\eta_{\mu_{j_1}\rho_{p_1}}...\eta_{\mu_{i_{\lambda_{13}}}\rho_{p_{\lambda_{13}}}}$$
$$\times\eta_{\mu_{k_1}\sigma_{s_1}}...\eta_{\mu_{k_{\lambda_{14}}}\sigma_{s_{\lambda_{14}}}}\eta_{\nu_{m_1}\rho_{r_1}}...\eta_{\nu_{m_{\lambda_{23}}}}\rho_{r_{\lambda_{23}}}$$
$$\eta_{\nu_{n_1}\sigma_{t_1}}...\eta_{\nu_{n_{\lambda_{24}}}}\rho_{t_{\lambda_{24}}}\eta_{\rho_{r_1}\sigma_{u_1}}...\eta_{\rho_{r_{\lambda_{34}}}}\sigma_{u_{\lambda_{34}}}$$
$$(i_1+l_1-2)!...(i_{\lambda_{12}}+l_{\lambda_{12}}-2)!$$
$$(j_1+p_1-2)!...(j_{\lambda_{13}}+p_{\lambda_{13}}-2)!$$
$$(k_1+s_1-2)!...(k_{\lambda_{14}}+s_{\lambda_{14}}-2)!$$
$$(m_1+q_1-2)!...(m_{\lambda_{23}}+q_{\lambda_{23}}-2)!$$
$$(n_1+t_1-2)!...(n_{\lambda_{24}}+t_{\lambda_{24}}-2)!$$
$$(r_1+u_1-2)!...(r_{\lambda_{34}}+u_{\lambda_{34}}-2)!$$
$$\times(-1)^{(s_1+\lambda_{12})s_2+s_3(s_4+\lambda_{34})+\lambda_{12}+\lambda_{34}+\lambda_{13}\lambda_{24}+\Delta_{23}+\Delta_{24}+\Delta_{34}}$$
$$(-1)^{\pi_q^{(s_1-3)}(i_1...i_{\lambda_{12}};j_1...j_{\lambda_{13}};k_1...k_{\lambda_{14}})}$$
$$(-1)^{\pi^{(s_2-2)}(l_1...l_{\lambda_{12}};m_1...m_{\lambda_{23}};n_1...n_{\lambda_{24}})}$$
$$(-1)^{\pi^{(s_3-2)}(p_1...p_{\lambda_{13}};q_1...q_{\lambda_{23}};r_1...r_{\lambda_{34}})}$$
$$(-1)^{\pi^{(s_4-2)}(s_1...s_{\lambda_{14}};t_1...t_{\lambda_{24}};u_1...u_{\lambda_{34}})} \tag{4.120}$$

where, by definition, the $\Xi^{(1-\psi)}_{\mu_1...\mu_{s_1-2}\nu_1...\nu_{s_2-2}\rho_1...\rho_{s_3-2}\sigma_1...\sigma_{s_4-2}}(\{\lambda\},\{\Delta\})$-factor, stemming from the summations over the orderings, which structure is explained below, is introduced in the equation above to abbreviate the notations. The sums in this formula are taken over the (non-ordered) partitions and over the orderings, with the notations (both in (4.120) and in the equations below) defined as follows:

1) $\sum_{n_1...n_p|\{\lambda_{ij}\}}$ denotes the sum over all the non-ordered partitions of $n_j>0; j=1,...p$ in non-negative $\lambda_{ij}=\lambda_{ji}$:

$$n_j=\sum_{k=1;k\neq j}^{p}\lambda_{jk} \tag{4.121}$$

Similarly, $\sum_{n_1...n_p|\{\Delta_{ij}\}}$ stands for the sum over all the non-ordered partitions of $n_j > 0; j = 1,...p$ in non-negative $\Delta_{ij}$ (in general, $\Delta_{ij} \neq \Delta_{ji}$)

$$n_j = \sum_{k=1;k\neq j}^{p} \Delta_{jk} \tag{4.122}$$

2) $\sum_{ord.\{k^{(1)}\}...\{k^{(p)}\}}^{(n_1;...n_p)}$ for each $n^{(j)}; j = 1,...p$ stands for the total $p$ summations over all the possible orderings of $n_j$ natural numbers from 1 to $n_j; 1 \leq j \leq p$ : $[k_1^{(j)}...k_{n_j}^{(j)}]$ $(1 \leq k_q^{(j)} \leq n_j; k_{q_1}^{(j)} \neq k_{q_2}^{(j)})$.

3) Similarly, for each $\sum_{ord.\{k^{(1)}\}_{q_1}...\{k_{q_j}^{(p)}\}}^{(n_1;...n_p)}$ stands for the total number of $p$ summations over the possible orderings of $n_j - 1$ natural numbers from 1 to $n_j$ with $q_j$ omitted.

4) $\pi^{(n)}(i_1,...i_n)$ stands for the number of nearest-neighbour permutations of a pair of numbers it takes to create the ordering $i_1...i_n$ of n numbers from 1 to $n$ from the ordering $1, 2, ...n$; $\pi_q^{(n)}(i_1,...i_{n-1})$ stands for the number of nearest-neighbour permutations of a pair of numbers it takes to creatye the ordering $i_1...i_{n-1}$ of $n - 1$ numbers from 1 to $n$ with $q$ omitted from the ordering $1, 2, ..., q - 1, q + 1, ...n$. The $\{i, j, k, l, m, n, p, q, r, s, t, u\}$ elements of the orderings in the formula satisfy the constraints:

$$\begin{aligned}
&i_1 + ... + i_{\lambda_{12}} = \delta_{12} \\
&j_1 + ... + j_{\lambda_{13}} = \delta_{13} \\
&k_1 + ... + k_{\lambda_{14}} = \delta_{14} \\
&l_1 + ... + l_{\lambda_{12}} = \delta_{21} \\
&m_1 + ... + m_{\lambda_{23}} = \delta_{23} \\
&n_1 + ... + j_{\lambda_{24}} = \delta_{24} \\
&p_1 + ... + p_{\lambda_{13}} = \delta_{31} \\
&q_1 + ... + q_{\lambda_{23}} = \delta_{32} \\
&r_1 + ... + r_{\lambda_{34}} = \delta_{34} \\
&s_1 + ... + s_{\lambda_{14}} = \delta_{41} \\
&t_1 + ... + t_{\lambda_{24}} = \delta_{42} \\
&u_1 + ... + u_{\lambda_{34}} = \delta_{43} \\
&\sum_j \delta_{1j} = (s_1 - 2)(s_1 - 3) - q \\
&\sum_j \delta_{ij} = (s_i - 2)(s_i - 3)
\end{aligned} \tag{4.123}$$

The partitions described above have a simple meaning in terms of the contractions between the $\psi$-fields contributing to the correlator (4.116). For each term, $\lambda_{ij} = \lambda_{ji}$ is the number of contractions between the higher-spin operators $V_{s_i}$ and $V_{s_j}$ $(i, j = 1, ..., 4; i \neq j)$. Obviously, for each $i$ fixed the sum of $\lambda_{ij}$ over $j$ gives the total number of $\psi$-fields in the $V_{s_i}$ vertex operator (equal to $s_i - 2$ for the operators in negative cohomologies and $s_i - 3$ or $s_i - 1$ for the operators in positive cohomologies upon the $K$-homotopy transformation). Next, for each term, $h_{ij} = \Delta_{ij} + \frac{1}{2}\lambda_{ij}$ is the conformal dimension that $V_{s_i}$ operator contributes to contractions with $V_{s_j}$, with $\Delta_{ij}$ stemming from the contributions from the derivatives and $\frac{1}{2}\lambda_{ij}$ from the total number $\lambda_{ij}$ of the $\psi$-fields themselves. The sum of $h_{ij}$ over $j$ obviously equals to the total conformal dimension of the $\psi$-fields and their derivatives in $V_{s_i}$ and is equal to $\frac{(s_j-2)^2}{2}$. Finally, the $(-1)^{\cdots}$-factors , including those of $(-1)^{\pi(\ldots)}$ stem from permutations of the $\psi$-fields participating in the contractions, as these fields have odd statistics. The second $\psi$-pattern, contributing to the overall correlator of the higher-spin operators i given by:

$$
\begin{aligned}
&A_\psi^{(2)} =<: \partial^r \psi_{\alpha_1} \{T^\perp \psi_{\mu(s_1-2)}\} : (\zeta_1) \{T^\perp \psi_{\nu(s_2-2)}\}(\zeta_2) \\
&\{T^\perp \psi_{\rho(s_3-2)}\}(0) \{T^\perp \psi_{\sigma(s_4-2)}\}(w \to \infty) > \\
&= \sum_{\{s_1-1, s_2-2, s_3-2, s_4-2 | \{\lambda_{ij}\}\}} \\
&\sum_{\{\frac{(s_1-2)(s_1-3)}{2}+r; \frac{(s_2-2)(s_2-3)}{2}; \frac{(s_3-2)(s_3-3)}{2}; \frac{(s_4-2)(s_4-3)}{2} | \{\Delta_{ij}\}\}} \\
&\{(\zeta_1 - \zeta_2)^{-\Delta_{12}-\Delta_{21}-\lambda_{12}} \zeta_1^{-\Delta_{13}-\Delta_{31}-\lambda_{13}} \\
&\times \zeta_2^{-\Delta_{23}-\Delta_{32}-\lambda_{23}} w^{-\frac{(s_4-2)(s_4-1)}{2}-\Delta_{41}-\Delta_{42}-\Delta_{43}} \\
&\sum_{ord.\{i,j,k\};\{l,m,n\};\{p,q,r\};\{s,t,u\}}^{(s_1-2)\oplus r; s_2-2; s_3-2; s_4-2} \eta_{\mu_{i_1} \nu_{l_1}} \ldots \eta_{\mu_{i_{\lambda_{12}}} \nu_{l_{\lambda_{12}}}} \\
&\times \eta_{\mu_{j_1} \rho_{p_1}} \ldots \eta_{\mu_{i_{\lambda_{13}}} \rho_{p_{\lambda_{13}}}} \\
&\eta_{\mu_{k_1} \sigma_{s_1}} \ldots \eta_{\mu_{k_{\lambda_{14}}} \sigma_{s_{\lambda_{14}}}} \eta_{\nu_{m_1} \rho_{r_1}} \ldots \eta_{\nu_{m_{\lambda_{23}}} \rho_{r_{\lambda_{23}}}} \\
&\eta_{\nu_{n_1} \sigma_{t_1}} \ldots \eta_{\nu_{n_{\lambda_{24}}} \rho_{t_{\lambda_{24}}}} \eta_{\rho_{r_1} \sigma_{u_1}} \ldots \eta_{\rho_{r_{\lambda_{34}}}} \sigma_{u_{\lambda_{34}}} \\
&(i_1 + l_1 - 2)! \ldots (i_{\lambda_{12}} + l_{\lambda_{12}} - 2)! \\
&(j_1 + p_1 - 2)! \ldots (j_{\lambda_{13}} + p_{\lambda_{13}} - 2)! \\
&(k_1 + s_1 - 2)! \ldots (k_{\lambda_{14}} + s_{\lambda_{14}} - 2)! \times
\end{aligned}
$$

$$
\begin{aligned}
&(m_1+q_1-2)!...(m_{\lambda_{23}}+q_{\lambda_{23}}-2)!\\
&(n_1+t_1-2)!...(n_{\lambda_{24}}+t_{\lambda_{24}}-2)!\\
&(r_1+u_1-2)!...(r_{\lambda_{34}}+u_{\lambda_{34}}-2)!\\
&\times(-1)^{s_1+(s_1+\lambda_{12})s_2+s_3(s_4+\lambda_{34})+\lambda_{12}+\lambda_{34}+\lambda_{13}\lambda_{24}+\Delta_{23}+\Delta_{24}+\Delta_{34}}\\
&(-1)^{\pi^{(s_1-1);r}(i_1...i_{\lambda_{12}};j_1...j_{\lambda_{13}};k_1...k_{\lambda_{14}})}\\
&(-1)^{\pi^{(s_2-2)}(l_1...l_{\lambda_{12}};m_1...m_{\lambda_{23}};n_1...n_{\lambda_{24}})}\\
&\times(-1)^{\pi^{(s_3-2)}(p_1...p_{\lambda_{13}};q_1...q_{\lambda_{23}};r_1...r_{\lambda_{34}})}\\
&(-1)^{\pi^{(s_4-2)}(s_1...s_{\lambda_{14}};t_1...t_{\lambda_{24}};u_1...u_{\lambda_{34}})}\\
&\equiv\sum_{s_1,s_2,s_3,s_4|\{\lambda_{ij};\Delta_{ij}\}}^{(2-\psi)}\Xi^{(2-\psi)}_{\mu(s_1-2)\nu(s_2-2)\rho(s_3-2)\sigma(s_4-2)}\\
&\times\frac{1}{(\zeta_1-\zeta_2)^{\Delta_{12}+\delta_{21}+\lambda_{12}}\zeta_1^{\Delta_{13}+\Delta_{31}+\lambda_{13}}\zeta_2^{\Delta_{23}+\Delta_{32}+\lambda_{23}}}\\
&\times w^{-(\frac{(s_4-2)(s_4-1)}{2}+\Delta_{41}+\Delta_{42}+\Delta_{43})}
\end{aligned}
\tag{4.124}
$$

where

$$\Xi^{(2-\psi)}_{\mu(s_1-2|r)\nu(s_2-2)\rho(s_3-2)\sigma(s_4-2)}(s_1,s_2,s_3,s_4|\{\lambda_{ij};\Delta_{ij}\})$$

with

$$\mu(s)=\prod_{k=1}^{s}\mu_k$$

and

$$\mu(s|r)\equiv\mu_1...\mu_{r-1}\mu_{r+1}...\mu_s;\, s\geq r$$

is again by definition given by the sum over the orderings defined in the previous equation, and the summation $\Sigma^{(2-\psi)}$ over the partitions of $s_i$ in $\lambda$'s and $\Delta$'s is as explained below. That is, the notations are the same as for the pattern 1 and, in addition, $\sum_{ord.\{k\}}^{n\oplus r}$ denotes the summation over the possible orderings $(k_1,...k_{n+1}$ of $n+1$ numbers from 1 to $n$ and $r>n$: (1,2,...,n,r) and $\pi^{n;r}(k_1,...k_{n+1}$ standing for the number of nearest-neighbor pair permutations it takes to make the ordering $(k_1,...k_{n+1})$ of $n+1$ numbers out of $(1,2,...,n,r)$. As in the pattern 1, the $\{i,j,k,l,m,n,p,q,r,s,t,u\}$ elements of the orderings are taken to satisfy the constraints:

$$
\begin{aligned}
&i_1+...+i_{\lambda_{12}}=\delta_{12}\\
&j_1+...+j_{\lambda_{13}}=\delta_{13}
\end{aligned}
$$

$$
\begin{aligned}
&k_1+...+k_{\lambda_{14}}=\delta_{14}\\
&l_1+...+l_{\lambda_{12}}=\delta_{21}\\
&m_1+...+m_{\lambda_{23}}=\delta_{23}\\
&n_1+...+j_{\lambda_{24}}=\delta_{24}\\
&p_1+...+p_{\lambda_{13}}=\delta_{31}\\
&q_1+...+q_{\lambda_{23}}=\delta_{32}\\
&r_1+...+r_{\lambda_{34}}=\delta_{34}\\
&s_1+...+s_{\lambda_{14}}=\delta_{41}\\
&t_1+...+t_{\lambda_{24}}=\delta_{42}\\
&u_1+...+u_{\lambda_{34}}=\delta_{43}\\
&\sum_j\delta_{1j}=(s_1-2)(s_1-3)+r\\
&\sum_j\delta_{ij}=(s_i-2)(s_i-3)
\end{aligned}
\tag{4.125}
$$

This concludes the computation of the $\psi$-factor of the 4-point correlator. Next, the $X$-factor of (4.116), straightforward to compute as well, is again given by the sum over the following partitions:

$$
\begin{aligned}
&A_X(n_1...n_4|p_1...p_4)\\
&\equiv <\prod_{i}^{n_1}\partial X_{\alpha_i}e^{ip_1X}(\zeta_1)\prod_{j=1}^{n_2}\partial X_{\beta_j}e^{ip_2X}(\zeta_2)\\
&\prod_{k=1}^{n_3}\partial X_{\gamma_k}e^{ip_3X}(0)\prod_{l=1}^{n_4}\partial X_{\delta_l}e^{ip_4X}(w\to\infty)>\\
&=\sum_{s_1-1,s_2-1,s_3-1,s_4-1|\{\kappa_{IJ}\};\{\tau_{IJ}\}}\frac{(s_1-1)!...(s_4-1)!}{\prod_{1\leq I,J\leq 4;I<J}\kappa_{IJ}!\tau_{IJ}!\tau_{JI}!}\\
&\times(\zeta_1-\zeta_2)^{-2\kappa_{12}-\tau_{12}-\tau_{21}}\zeta_1^{-2\kappa_{13}-\tau_{13}-\tau_{31}}\zeta_2^{-2\kappa_{23}-\tau_{23}-\tau_{32}}\\
&\times w^{-s_4+1-\kappa_{41}-\kappa_{42}-\kappa 43-\tau_{41}-\tau_{42}-\tau_{43}}\\
&\times\prod_{\{1\leq k_1\leq\kappa_{12}\}}\prod_{\{\kappa_{12}+1\leq k_2\leq\kappa_{12}+\kappa_{13}\}}\prod_{\{\kappa_{12}+\kappa_{13}+1\leq k_3\leq\kappa_{12}+\kappa_{13}+\kappa_{14}\}}\\
&\prod_{\{\kappa_{12}+\kappa_{13}+\kappa_{14}+1\leq k_4\leq\kappa_{12}+\kappa_{13}+\kappa_{14}+\tau_{12}\}}\\
&\prod_{\{\kappa_{12}+\kappa_{13}+\kappa_{14}+\tau_{12}+1\leq k_5\leq\kappa_{12}+\kappa_{13}+\kappa_{14}+\tau_{12}+\tau_{13}\}}
\end{aligned}
$$

$$\prod_{\{\kappa_{12}+\kappa_{13}+\kappa_{14}+\tau_{12}+\tau_{13}+1\leq k_6\leq s_1-1\}}\prod_{\{1\leq l_1\leq\kappa_{12}\}}$$

$$\prod_{\{\kappa_{12}+1\leq l_2\leq\kappa_{12}+\kappa_{23}\}}\prod_{\{\kappa_{12}+\kappa_{23}+1\leq l_3\leq\kappa_{12}+\kappa_{23}+\kappa_{24}\}}$$

$$\prod_{\{\kappa_{12}+\kappa_{23}+\kappa_{24}+1\leq l_4\leq\kappa_{12}+\kappa_{23}+\kappa_{24}+\tau_{21}\}}$$

$$\prod_{\{\kappa_{12}+\kappa_{23}+\kappa_{24}+\tau_{21}+1\leq l_5\leq\kappa_{12}+\kappa_{23}+\kappa_{24}+\tau_{21}+\tau_{23}\}}$$

$$\prod_{\{\kappa_{12}+\kappa_{23}+\kappa_{24}+\tau_{21}+\tau_{23}+1\leq l_6\leq s_2-1\}}\prod_{\{1\leq m_1\leq\kappa_{13}\}}\prod_{\{\kappa_{13}+1\leq m_2\leq\kappa_{13}+\kappa_{23}\}}$$

$$\prod_{\kappa_{13}+\kappa_{23}+1\leq m_3\leq\kappa_{13}+\kappa_{23}+\kappa_{34}}$$

$$\prod_{\kappa_{13}+\kappa_{23}+\kappa_{34}+1\leq m_4\leq\kappa_{13}+\kappa_{23}+\kappa_{34}+\tau_{31}}$$

$$\prod_{\{\kappa_{13}+\kappa_{23}+\kappa_{34}+\tau_{31}+1\leq m_5\leq\kappa_{13}+\kappa_{23}+\kappa_{34}+\tau_{31}+\tau_{32}\}}$$

$$\prod_{\{\kappa_{12}+\kappa_{23}+\kappa_{34}+\tau_{31}+\tau_{32}+1\leq m_6\leq s_3-1\}}$$

$$\prod_{\{1\leq n_1\leq\kappa_{14}\}}\prod_{\{\kappa_{14}+1\leq n_2\leq\kappa_{14}+\kappa_{24}\}}\prod_{\{\kappa_{14}+\kappa_{24}+1\leq n_3\leq\kappa_{14}+\kappa_{24}+\kappa_{34}\}}$$

$$\prod_{\{\kappa_{14}+\kappa_{24}+\kappa_{34}+1\leq n_4\leq\kappa_{14}+\kappa_{24}+\kappa_{34}+\tau_{41}\}}$$

$$\prod_{\{\kappa_{14}+\kappa_{24}+\kappa_{34}+\tau_{41}+1\leq n_5\leq\kappa_{14}+\kappa_{24}+\kappa_{34}+\tau_{41}+\tau_{42}\}}$$

$$\prod_{\{\kappa_{14}+\kappa_{24}+\kappa_{34}+\tau_{41}+\tau_{42}+1\leq n_6\leq s_4-1\}}$$

$$\times\eta_{\alpha_{k_1}\beta_{l_1}}\eta_{\alpha_{k_2}\gamma_{m_1}}\eta_{\alpha_{k_3}\delta_{n_1}}\eta_{\beta_{l_2}\gamma_{m_2}}\eta_{\beta_{l_3}\delta_{n_2}}$$

$$\eta_{\gamma_{m_3}\delta_{n_3}}(ip_2)_{\alpha_{k_4}}(ip_3)_{\alpha_{k_5}}(ip_4)_{\alpha_{k_6}}(-ip_1)_{\beta_{l_4}}(ip_3)_{\beta_{l_5}}(ip_4)_{\beta_{l_6}}$$

$$(-ip_1)_{\gamma_{m_4}}(-ip_2)_{\gamma_{m_5}}(ip_4)_{\gamma_{m_6}}(-ip_1)_{\delta_{n_4}}(-ip_2)_{\delta_{n_5}}(-ip_3)_{\delta_{n_6}}$$

$$\times\delta(p_1+p_2+p_3+p_4)$$

$$\equiv \sum_{s_1-1,s_2-1,s_3-1,s_4-1|\{\kappa_{IJ}\};\{\tau_{IJ}\}} \frac{(s_1-1)!...(s_4-1)!}{\prod_{1\leq I,J\leq 4;I<J}\kappa_{IJ}!\tau_{IJ}!\tau_{JI}!}$$
$$(\zeta_1-\zeta_2)^{-2\kappa_{12}-\tau_{12}-\tau_{21}}\zeta_1^{-2\kappa_{13}-\tau_{13}-\tau_{31}}$$
$$\times\zeta_2^{-2\kappa_{23}-\tau_{23}-\tau_{32}}w^{-s_4+1-\kappa_{41}-\kappa_{42}-\kappa 43-\tau_{41}-\tau_{42}-\tau_{43}}$$
$$\times\Xi^{(X)}_{\alpha(s_1-1)\beta(s_2-1)\gamma(s_3-1)\delta(s_4-1)}(s_1-1,...s_4-1|\{\kappa\};\{\tau\}) \tag{4.126}$$

where the factors $\Xi^{(X)}(n_1...n_4|\{\kappa\};\{\tau\})$ are by definition introduced according to (4.126) and the sum $\sum_{n_1,n_2,n_3,n_4|\{\kappa_{IJ}\};\{\tau_{IJ}\}}$ is taken over the non-ordered partitions:

$$n_I = \sum_{1\leq J\leq 4;J\neq I}(\kappa_{IJ}+\tau_{IJ})$$
$$\kappa_{IJ}=\kappa_{JI} \tag{4.127}$$

The final ingredient for the higher-spin amplitude comes from the ghost factor. First of all, note that, since the Bell polynomials in the ghost fields appearing in the expressions for the vertex operators are limited to $B^{(m)}_{\phi-\chi}$ and $B^{(n)}_{2\phi-2\chi-\sigma}$ and the operator products between the derivatives of $\phi-\chi$ and $2\phi-2\chi-\sigma$ are nonsingular, the polynomials only contract with the ghost exponentials. The pattern for the ghost part of the correlator is given by

$$A_{gh}(s_1...s_4|n)$$
$$=<: ce^{\chi+(s_1-3)\phi}B^{(n)}_{\phi-\chi}:(\zeta_1):e^{(s_2-2)\phi}B^{(2s_2-4)}_{2\phi-2\chi-\sigma}:(\zeta_2)$$
$$ce^{-s_3\phi}(0)ce^{-s_4\phi}(w\to\infty)>$$
$$=\sum_{\{n|\omega_{12},\omega_{13},\omega_{14}\}}\sum_{\{2s_2-4|\omega_{21},\omega_{23},\omega_{24}\}}\Xi^{(gh)}(s_1...s_4|n;\{\omega\})$$
$$\times(\zeta_1-\zeta_2)^{-(s_1-3)(s_2-2)-\omega_{12}-\omega_{21}}\zeta_1^{(s_1-3)s_3+1-\omega_{13}}$$
$$\times\zeta_2^{(s_2-2)s_3+1-\omega_{23}}w^{(s_1+s_2-5)s_4+2-\omega_{14}-\omega_{24}} \tag{4.128}$$

where

$$\Xi^{(gh)}(s_1...s_4|n;\{\omega\})=C(2-s_2|n;n-\omega_{12})$$
$$\times C(s_3|n-\omega_{12};n-\omega_{12}-\omega_{13})$$
$$\times C(s_4|n-\omega_{12};n-\omega_{12}-\omega_{13};0)$$
$$\times C(3-2s_1|2s_2-4;2s_2-4-\omega_{21})$$
$$\times C(2s_3-1|2s_2-4-\omega_{21};2s_2-4-\omega_{21}-\omega_{23})$$
$$\times C(2s_4-1|2s_2-4-\omega_{12}-\omega_{23};0) \tag{4.129}$$

where the coefficients $C(q|n; n-m)$ stem from the operator product:

$$B^{(N)}_{\alpha_1\phi+\alpha_2\chi+\alpha_3\sigma}(z)e^{\beta_1\phi+\beta_2\chi+\beta_3\sigma}(w)$$
$$=\sum_{n=0}^{N}(z-w)^{-n}C(-\alpha_1\beta_1+\alpha_2\beta_2+\alpha_3\beta_3|N;N-n)$$
$$:B^{(N-n)}_{\alpha_1\phi+\alpha_2\chi+\alpha_3\sigma}(z)e^{\beta_1\phi+\beta_2\chi+\beta_3\sigma}:(w) \tag{4.130}$$

with

$$C(a|N;N-n)=\frac{\Gamma(1+a)}{n!\Gamma(1+a-n)} \tag{4.131}$$

and $\omega_{ij}$-numbers satisfy the constraints

$$\omega_{12}+\omega_{13}+\omega_{14}=n$$
$$\omega_{21}+\omega_{23}+\omega_{24}=2s_2-4 \tag{4.132}$$

and define the non-ordered partitions of $n$ and $2s_2-4$, with the summations taken over these partitions. This concludes the computation of all of the patterns ($\psi$,$X$, and ghost) contributing to the 4-point correlation function of the higher spins. Finally, using

$$\int_0^1 dz_1\int_0^{z_1} dz_2{z_1}^a z_2^b(z_1-z_2)^c=\frac{\Gamma(a+1)\Gamma(c+1)}{(a+b+c+2)\Gamma(a+c+2)} \tag{4.133}$$

substituting the patterns into the integrals in (4.116) and integrating, we obtain the following answer for the amplitude:

$$A(s_1,s_2,s_3,s_4)=\frac{\Omega^{s_1|s_1-3}(p_1)\Omega^{s_2|s_2-3}(p_2)\Omega^{s_3|s_3-3}(p_3)\Omega^{s_4|s_4-3}(p_4)}{8}$$
$$\times\{\sum_{q=1}^{s_1-2}(-1)^q(q-1)!$$
$$\sum_{\{2s_2-4|\omega_{21},\omega_{23},\omega_{24}\}}\ \sum_{\{s_1+q+2|\omega_{12},\omega_{13},\omega_{14}\}}\ \sum_{\{s_1-3,s_2-2,s_3-2,s_4-2|\{\lambda_{ij}\}\}}$$
$$\sum_{\frac{(s_1-2)(s_1-3)}{2}-q;\frac{(s_2-2)(s_2-3)}{2};\frac{(s_3-2)(s_2-3)}{2};\frac{(s_4-2)(s_4-3)}{2}|\{|\Delta_{ij}\}}$$
$$(\sum_{s_1-2;s_2-1;s_3-1;s_4-1|\{\tau\};\{\kappa\}}\frac{(s_1-1)!...(s_4-1)!}{\prod_{1\leq I,J\leq 4;I<J}\kappa_{IJ}!\tau_{IJ}!\tau_{JI}!}$$
$$\Xi^{(1-\psi)}_{\mu(s_1-2;q)\nu(s_2-2)\rho(s_3-2)\sigma(s_4-2)}(\{\lambda\},\{\Delta\})$$
$$(-\eta_{\mu_q\alpha_q}\Xi^{(X)}_{\alpha(s_1-1|q)\beta(s_2-1)\gamma(s_3-1)\delta(s_4-1)}(s_1-1,...s_4-1|\{\kappa\};\{\tau\})$$

$$-\sum_{s_1-1;s_2-1;s_3-1;s_4-1|\{\tau\};\{\kappa\}}$$

$$(ip_1)_{\mu_q}\Xi^{(X)}_{\alpha(s_1-1)\beta(s_2-1)\gamma(s_3-1)\delta(s_4-1)}(s_1-1,...s_4-1|\{\kappa\};\{\tau\})$$

$$\Xi^{(gh)}(s_1...s_4|s_1+q+2;\{\omega\})F(p_1...p_4|\{\lambda,\Delta,\kappa,\tau,\omega\}))$$

$$+\sum_{q=1}^{s_1-2}\sum_{r=0}^{s_1+q}(-1)^q(q-1)!$$

$$\sum_{\{2s_2-4|\omega_{21},\omega_{23},\omega_{24}\}}\sum_{\{s_1+q-r|\omega_{12},\omega_{13},\omega_{14}\}}\sum_{\{s_1-3,s_2-2,s_3-2,s_4-2|\{\lambda_{ij}\}\}}$$

$$\sum_{\{\frac{(s_1-2)(s_1-3)}{2}-q;\frac{(s_2-2)(s_2-3)}{2};\frac{(s_3-2)(s_2-3)}{2};\frac{(s_4-2)(s_4-3)}{2}|\{|\Delta_{ij}\}\}}$$

$$\times(\sum_{s_1-2;s_2-1;s_3-1;s_4-1|\{\tau\};\{\kappa\}}(-1)^q(q-1)!$$

$$\frac{(s_1-1)!...(s_4-1)!}{\prod_{1\leq I,J\leq 4;I<J}\kappa_{IJ}!\tau_{IJ}!\tau_{JI}!}\Xi^{(1-\psi)}_{\mu(s_1-2|q)\nu(s_2-2)\rho(s_3-2)\sigma(s_4-2)}(\{\lambda\},\{\Delta\})$$

$$\times\Xi^{(gh)}(s_1...s_4|s_1+q-r;\{\omega\})$$

$$\times[((-1)^{r+1}\Xi^{(X)}_{\alpha(s_1-1)\beta(s_2-1|q)\gamma(s_3-1)\delta(s_4-1)}(s_1-1,...s_4-1|\{\kappa\};\{\tau\})$$

$$\times(\eta_{\alpha_q\beta_q}(r+1)!F^{(1)}_r(p_1...p_4|\{\lambda,\Delta,\kappa,\tau,\omega\})$$

$$+(-1)^{r+1}\Xi^{(X)}_{\alpha(s_1-1)\beta(s_2-1)\gamma(s_3)\delta(s_4-1)}(s_1-1,...s_4-1|\{\kappa\};$$

$$\{\tau\})(ip_2)^q r!$$

$$\times F^{(1)}_{r-1}(p_1...p_4|\{\lambda,\Delta,\kappa,\tau,\omega\})(s_1-1,...s_4-1|\{\kappa\};\{\tau\})))$$

$$+((-1)^{r+1}\Xi^{(X)}_{\alpha(s_1-1)\beta(s_2-1)\gamma(s_3-1|q)\delta(s_4-1)}(s_1-1,...s_4-1|\{\kappa\};\{\tau\})$$

$$\times(\eta_{\alpha_q\gamma_q}(r+1)!F^{(2)}_r(p_1...p_4|\{\lambda,\Delta,\kappa,\tau,\omega\})$$

$$+(-1)^{r+1}\Xi^{(X)}_{\alpha(s_1-1)\beta(s_2-1)\gamma(s_3)\delta(s_4-1)}(s_1-1,...s_4-1|\{\kappa\};\{\tau\})$$

$$\times(ip_3)^q r!F^{(2)}_{r-1}(p_1...p_4|\{\lambda,\Delta,\kappa,\tau,\omega\})(s_1-1,...s_4-1|\{\kappa\};\{\tau\})))$$

$$+((-1)^{r+1}\Xi^{(X)}_{\alpha(s_1-1)\beta(s_2-1)\gamma(s_3-1)\delta(s_4-1|q)}$$

$$\times(\eta_{\alpha_q\delta_q}(r+1)!F^{(3)}_r(p_1...p_4|\{\lambda,\Delta,\kappa,\tau,\omega\})$$

$$+(-1)^{r+1}\Xi^{(X)}_{\alpha(s_1-1)\beta(s_2-1)\gamma(s_3)\delta(s_4-1)}(s_1-1,...s_4-1|\{\kappa\};\{\tau\})$$

$$\times(ip_4)^q r!F^{(3)}_{r-1}(p_1...p_4|\{\lambda,\Delta,\kappa,\tau,\omega\}))))]$$

$$
\begin{aligned}
&-\sum_{r=s_1-1}^{s_1+2}\sum_{\{s_1+2-r|\omega_{12},\omega_{13},\omega_{14}\}}\\
&\sum_{\{2s_2-4|\omega_{21},\omega_{23},\omega_{24}\}}\sum_{\{s_1+q+2|\omega_{12},\omega_{13},\omega_{14}\}}\\
&\sum_{\{s_1-1,s_2-2,s_3-2,s_4-2|\{\lambda_{ij}\}\}}\\
&\sum_{\{\frac{(s_1-2)(s_1-3)}{2}+r;\frac{(s_2-2)(s_2-3)}{2};\frac{(s_3-2)(s_2-3)}{2};\frac{(s_4-2)(s_4-3)}{2}|\{|\Delta_{ij}\}\}}\\
&\frac{1}{r!}\frac{(s_1-1)!...(s_4-1)!}{\prod_{1\le I,J\le 4;I<J}\kappa_{IJ}!\tau_{IJ}!\tau_{JI}!}\\
&\times\Xi^{(2-\psi)}_{\mu_r|\mu(s_1-2)\nu(s_2-2)\rho(s_3-2)\sigma(s_4-2)}(s_1,s_2,s_3,s_4|\{\lambda_{ij};\Delta_{ij}\})\\
&\times\Xi^{(X)}_{\alpha(s_1-1|r)\beta(s_2-1)\gamma(s_3-1)\delta(s_4-1)}(s_1-1,...s_4-1|\{\kappa\};\{\tau\})\\
&\times\Xi^{(gh)}(s_1...s_4|s_1+2-r;\{\omega\})F(p_1...p_4|\{\lambda,\Delta,\kappa,\tau,\omega\})\\
&-\sum_{r=s_1-1}^{s_1+1}\sum_{\{n|\omega_{12},\omega_{13},\omega_{14}\}}\sum_{\{2s_2-4|\omega_{21},\omega_{23},\omega_{24}\}}\\
&\sum_{\{s_1+1-r|\omega_{12},\omega_{13},\omega_{14}\}}\sum_{\{s_1-3,s_2-2,s_3-2,s_4-2|\{\lambda_{ij}\}\}}\\
&\sum_{\{\frac{(s_1-2)(s_1-3)}{2}+r;\frac{(s_2-2)(s_2-3)}{2};\frac{(s_3-2)(s_2-3)}{2};\frac{(s_4-2)(s_4-3)}{2}|\{|\Delta_{ij}\}\}}\\
&\frac{1}{r!}\frac{(s_1-1)!...(s_4-1)!}{\prod_{1\le I,J\le 4;I<J}\kappa_{IJ}!\tau_{IJ}!\tau_{JI}!}(ip^{\mu_r})\\
&\times\Xi^{(2-\psi)}_{\mu_r;\mu(s_1-2)\nu(s_2-2)\rho(s_3-2)\sigma(s_4-2)}(s_1,s_2,s_3,s_4|\{\lambda_{ij};\Delta_{ij}\})\\
&\times\Xi^{(X)}_{\alpha(s_1-1)\beta(s_2-1)\gamma(s_3-1)\delta(s_4-1)}(s_1-1,...s_4-1|\{\kappa\};\{\tau\})\\
&\times\Xi^{(gh)}(s_1...s_4|s_1+1-r;\{\omega\})F(p_1...p_4|\{\lambda,\Delta,\kappa,\tau,\omega\})
\end{aligned}
\tag{4.134}
$$

where

$$
\begin{aligned}
&F(p_1...p_4|\{\lambda,\Delta,\kappa,\tau,\omega\})\equiv F(A;B;C)\\
&=\frac{\Gamma(1+C(\{\lambda,\Delta,\tau,\gamma,\omega,\{s\}\}))\Gamma(1+A(\{\lambda,\Delta,\tau,\gamma,\omega,\{s\}\}))}{(A+B+C+2)\Gamma(A+C+2)}
\end{aligned}
\tag{4.135}
$$

with

$$C = p_1p_2 - \Delta_{12} - \Delta_{21} - \lambda_{12} - \tau_{12} - \tau_{21}$$
$$-2\kappa_{12} - (s_1 - 3)(s_2 - 2) - \omega_{12} - \omega_{21}$$
$$A = p_1p_3 - \Delta_{13} - \Delta_{31} - \lambda_{13} - \tau_{13} - \tau_{31}$$
$$-2\kappa_{13} + (s_1 - 3)s_3 - \omega_{13}$$
$$B = -\frac{1}{2}(p_1p_4 + p_2^2 + p_3^2 + p_4^2) - \omega_{23}$$
$$-\Delta_{23} - \Delta_{32} - \lambda_{23} - \tau_{23} - \tau_{32} - 2\kappa_{23} + (s_2 - 2)s_3 + 1) \tag{4.136}$$

and

$$F_r^{(1)}(A; B; C) = F(A; B; C + r)$$
$$F_r^{(2)}(A; B; C) = F(A + r; B; C)$$
$$F_r^{(3)}(A; B; C) = F(A; B + r; C) \tag{4.137}$$

Finally, all the summations over the partitions in the higher-spin amplitude are subject to one more unitarity constraint:

$$s_4(s_1 + s_2 - 6) + 3 - \frac{1}{2}(s_4 - 2)(s_4 - 1)$$
$$-\sum_{j=1}^{3}(\Delta_{4j} - \kappa_{4j} - \tau_{4j}) - \omega_{14} - \omega_{24} = 0 \tag{4.138}$$

which stems from the condition that, in the limit $w \to \infty$, as the location of the $s_4$ vertex operator taken to the infinity, only the terms behaving as $\sim w^0$ survive.

This concludes the derivation of the general 4-point amplitude for the higher spins in AdS, provided that the spin values satisfy the constraint (4.115) and the particles are polarized and propagating along the *AdS* boundary.

## 4.7 A Glance at Quintics

Our calculation of the 4-point function for the higher spins in AdS related the nonlocalities of the quartic interactions to the specific structure of the higher-spin vertex operators in string theory. In the calculation, we particularly used the relative simplicity of the vertex operators for the $\Omega^{s-1|t}$ frame-like fields, in the case $t = s - 3$.

For the $t$ values other than $t = s - 3$ the manifest expressions for the higher spin vertex operators become significantly more complicated.

As it was explained in the section 4.4, the explicit relation between the vertex operators $V_{s-1|t} \equiv \Omega^{s-1|t}W_{s-1|t}$ and $V_{s-1|s-3}$ is given by the generalized zero-torsion constraints (4.88), which become the operator identities in RNS string theory.

In particular, to obtain the operator for the Fronsdal's field one can take $k = s-3$ and $t = 0$. Note that all the $V_{s-1|t} \equiv \Omega^{s-1|t}W_{s-1|t}$ higher spin vertex operators are the elements of $H_{-s} \sim H_{s-2}$ cohomology for all the values of $0 \leq t \leq s-3$, although the canonical pictures for $V_{s-1|t}$ are different and equal to $-2s+t+3$ at the negative picture representation. The action of $\Gamma$ on $V_{s|t}$ thus results in increasing the ghost picture by one unit and the appearance of the extra $p$ factor in front of $\Omega(p)$, typically due to the contraction of the first term in $\Gamma$ with $e^{ipX}$ factor in the vertex operator. The zero torsion constraints (4.88) can of course be reformulated equivalently for the operators in the positive cohomology representations; will all the spin $s$ operators being the elements of $H_{s-2}$, the constraints for the frame-like vertex operators in the positive cohomologies are

$$\begin{aligned}
&\Omega^{s-1|t} : \Gamma^{-k}W_{s-1|t} := \Omega^{s-1|t+1}W_{s-1|t+k}(k \leq s-3-t)\\
&\Omega^{s-1|t} : \Gamma^{-k}W_{s-1|t} := 0(k > s-3-t)
\end{aligned} \tag{4.139}$$

with $\Omega^{s-1|t}W_{s-1|t}$-operators having canonical ghost pictures $2s-5-t$, in particular the operator for the Fronsdal's field having canonical positive picture $2s-5$. Here $: \Gamma^{-k} :=: (\Gamma^{-1})^k :$ with $\Gamma^{-1} = -4ce^{\chi-2\phi}\partial\chi$ being the *inverse* picture-changing operator.

In practice, the operator equations (4.88), (4.139) generating the generalized zero torsion constraints by the picture-changing relations, are hard to solve. For example, the solution for the simplest of the equations (4.88) for $t = s-4$ is already complicated enough, with the vertex operator having the form

$$\begin{aligned}
&V_{s-1|s-4}(p;z)\\
&\sim ce^{-(s+1)\phi}\partial X_{m_1}...\partial X_{s-1}\psi^{\alpha_0}\partial\psi_{\alpha_1}\partial^2\psi_{\alpha_2}...\partial^{s-4}\psi_{\alpha_{s-4}}\\
&\times[\alpha B_{-\phi}^{(2s-3)}\sum_{p=0}^{s-3}\sum_{q=p+1}^{s-2}\alpha_{p|q}\partial^p\psi_m\partial^q\psi^n B_{-\phi}^{(2s-4-p-q)}\\
&+\alpha_{s-3|s-1}\partial^{s-3}\psi_m\partial^{s-1}\psi^m]e^{ipX}
\end{aligned} \tag{4.140}$$

at canonical $-s-1$-picture where the $\alpha_{p|q}$-coefficients must be calculated so as to ensure that $V_{s-1|s-4}(p;z)$ is primary (i.e. the singularities of cubic and higher orders stemming from the OPE of the stress-tensor with the

$\psi$-part must be cancelled by those stemming from the operator product with the Bell polynomials in the derivatives of the $\phi$-ghost field). Note that the leading order of the operator products of all the Bell polynomials $B_{-\phi}^{(N)}$ with $e^{\phi}$ is the simple pole for any $N$, which ensures that the picture-changing transformation of (17) does not produce any terms other than those proportional to $V_{s-1|s-3}$, as well as the absence of singularities in the OPE of $\Gamma$ and $V_{s-1|s-4}(p;z)$. It is now not hard to see that the expressions for operators with $t \leq s-5$ will get more and more tedious for the lower $t$ values; their general structure would involve sums over multiple products of $\partial^p\vec{\psi}\partial^q\vec{\psi}$-factors multiplied by products of $s-3-t$ Bell polynomials of the ghost fields, making them cumbersome objects to work with.

Thus the $V_{s-1|s-3}$-operators appear to be particularly convenient and natural objects to use in order to describe the higher-spin vertices in space-time. However, there are restrictions on the spin values for the vertices that can be described by the correlators with all the operators being of the $V_{s-1|s-3}$-only. That is, for a $n = p+q$-point higher spin amplitude containing $p$ $V_{s-1|s-3}$-operators at positive cohomologies and $q$ operators at negative cohomologies the spin values must satisfy

$$\sum_{j=1}^{p}(s_j-2) - \sum_{j=p+1}^{p+q} s_j = -2 \tag{4.141}$$

in order to satisfy the superconformal ghost balance constraint. The amplitudes with spin values not satisfying (4.141) cannot be described by using solely the operators of this type and require finding explicit solutions of the operator equations (4.88), (4.139) making them far more complicated. At the same time, note that if the constraint (4.141) is satisfied, despite the fact that the correlation functions of the $V_{s-1|s-3}$-type operators by construction contain certain minimal number of derivatives (since each space-time field $\Omega_{s-1|s-3}$ by definition contains $s-3$ derivatives), all the amplitudes involving $V_{s-1|s-3}$ can be cast equivalently in terms of those involving operators for the Fronsdal's fields and/or the operators for the extra fields $\Omega_{s-1|t}$ with lower $t$, using the zero torsion relations, combined with the picture equivalence of the operators inside each particular cohomology. For example, consider a 5-point higher spin amplitude with the spins satisfying the constraints $s_1+s_2-s_3-s_4-s_5=2$. In the amplitude of this type, two operators are integrated at positive picture, and three operators are unintegrated at negative pictures. Using the zero torsion relations, we get

$$\begin{aligned}
&A(s_1,...s_5)\\
&=<\Omega^{s_1-1|s_1-3}W^{(s_1-2)}_{s_1-1|s_1-3}(p_1)\Omega^{s_2-1|s_2-3}W^{(s_2-2)}_{s_2-1s_2-3}(p_2)\\
&\Omega^{s_3-1|s_3-3}W^{(-s_3)}_{s_3-1|s_3-3}(p_3)\Omega^{s_4-1|s_4-3}W^{(-s_4)}_{s_4-1|s_4-3}(p_4)\\
&\Omega^{s_5-1|s_5-3}W^{(-s_5)}_{s_5-1|s_5-3}(p_5)>\\
&=<:\Gamma^{-s_1-s_2+6}:\Omega^{s_1-1|0}W^{(2s_1-5)}_{s_1-1|s_1-3}(p_1)\Omega^{s_2-1|0}W^{(2s_2-5)}_{s_2-1s_2-3}(p_2)\\
&:\Gamma^{s_3+s_4+s_5-9}:\Omega^{s_3-1|0}W^{(3-2s_3)}_{s_3-1|0}(p_3)\\
&\Omega^{s_4-1|0}W^{(3-2s_4)}_{s_4-1|0}(p_4)\Omega^{s_5-1|0}W^{(3-2s_5)}_{s_5-1|0}(p_5)>\\
&=<\Omega^{s_1-1|0}W^{(2s_1-6)}_{s_1-1|0}(p_1)\\
&\Omega^{s_2-1|0}W^{(2s_2-6)}_{s_2-1|0}(p_2)\Omega^{s_3-1|0}W^{(2-2s_3)}_{s_3-1|0}(p_3)\\
&\Omega^{s_4-1|0}W^{(2-2s_4)}_{s_4-1|0}(p_4)\Omega^{s_5-1|0}W^{(2-2s_5)}_{s_5-1|0}(p_5)>
\end{aligned}\tag{4.142}$$

i.e. the amplitude involving five $V_{s-1|s-3}$ higher spin operators is identical to the one involving five Fronsdal operators at pictures each lowered by one unit with respect to the canonical.

As it is clear from the above discussion, the overall $b-c$ and superconformal ghost structures of the $n=p+q$-point higher spin amplitudes, combined with the ghost structure of the picture-changing operators suggest the existence of two types of higher spin interactions at higher orders: the first type having $p=n-3, q=3$. This amplitude involves 3 unintegrated negative picture operators and standard $N-3$ integrated operators at positive pictures. All the terms, contributed by all the integrated operators, are of $A_0$-type (having $b-c$ ghost number zero) The higher spin amplitudes of this type have the standard Veneziano pole structure, leading to local interaction terms in the low-energy effective action (with the poles in the amplitudes corresponding to different channels of particle exchanges). These poles do not produce any physical nonlocalities; the nonlocalities that one may encounter upon the space-time momentum integration and expressing the low-energy effective action in the position space, are not physical and can be removed by substituting the $\beta$-function equations at lower orders.

The amplitudes of the second type, on the other hand, have the structure $p=n-2, q=2$.They involve two unintegrated operators contributing two $c$-ghosts, with the third $c$-ghost stemming from the $A_1$-type terms of one of the integrated operators and the remaining operators contributing $A_0$-type terms. These amplitudes have the structure very different from those of the first type. Due to the extra integration, they contain extra poles, corresponding to physical nonlocalities, rather than particle exchanges. Unlike the Veneziano-type case, the nonlocalities in the position space, obtained upon the Fourier transform, cannot be removed using the $\beta$-function flows, but reflect genuine nonlocalities of the higher spin interactions.

In fact, generically most of the higher spin amplitudes are of the second type, with the first type emerging only for the special combination of the spin values. We will refer to the appearance of the first type higher spin amplitudes as the "localization". In fact, as we shall point out below, such a localization effect does not occur at the quartic order but only appears at quintic and higher order interactions. Below we shall investigate this effect by direct computation of the correlation functions.

## 4.8 Quintic Interactions and Localization

We start with the 4-point higher spin amplitude with the spin values satisfying constraint (4.141) with $p = 1, q = 3$. It is not difficult to see that this 4-point amplitude vanishes. Indeed, the $V_{s_1}$ operator at positive cohomology contains $s_1 - 2 = s_1 + s_2 + s_3 - 2$ $\psi$-fields, which cannot fully contract to the RNS fermions of the remaining 3 operators, since the operators at negative pictures contribute altogether $s_1 + s_2 + s_3 - 6$ $\psi$-fields. This means that the 4-point amplitude with such spin values admits no localization. Next, consider the 5-point amplitude with the localization constraint $p = 2, q = 3$. To simplify the calculations as much as possible, we assume that the value one of the spins is small enough (namely, $s_4 \equiv u = 4$) and $s_5 + u - s_1 - s_2 - s_3 = 2$. All the vertex operators for the frame-like fields are in the $V_{s-1|s-3}$ representation, with the operators the $s_1, s_2, s_3$ being unintegrated at negative pictures and the operators for spins $u = 4$ and $s_5 = \sum s - 2$ at integrated positive (to abbreviate the notations, denote $\sum s = s_1 + s_2 + s_3$). With such a picture arrangement, only $A_0$-type terms of the both of the integrated operators contribute to the correlator. We are now all set to compute, step by step, the $X$,ghost and $\psi$-factors of the correlator defining the quintic interaction of the above spin values. We start with the calculation of the $X$-part. The correlation function is given by

$$
\begin{aligned}
&A_X(p_1, ...p_5|w; z; \xi) \\
&=< \partial X^{m_1} ... \partial X^{m_{s_1-1}} e^{ip_5 X}(w)|_{w\to\infty} \partial X^{n_1} ... \partial X^{n_{s_2-1}} e^{ip_4 X}(1) \\
&\partial X^{q_1} ... \partial X^{q_{u-1}} e^{ip_3 X}(z) \partial X^{r_1} ... \partial X^{r_{\Sigma s+1-u}} e^{ip_2 X}(\xi) \\
&\partial X^{t_1} ... \partial X^{t_{s_3-1}} e^{ip_1 X}(0) >
\end{aligned}
\tag{4.143}
$$

where we have arranged the operator's insertions in the order $(z_1 = w) > (z_2 = 1) > z > \xi > (z_3 = 0)$, with $z_{1,2,3}$ being the locations of the spin $s_{1,2,3}$ unintegrated vertices, $z$ and $\xi$ are the insertion points of the remaining spins (to be integrated over). We shall later set $z_3 \equiv w \to \infty$ but for now shall the $w$-dependence manifest to keep track of the infinities. To compute the correlator, it is useful to define the partitions

$$s_i - 1 = \sum_{j=1,j\neq i}^{5} (R_{ij} + Q_{ij}); i,j = 1,...,5 \tag{4.144}$$

where $R_{ij}$ is the number of contractions between $\partial X$'s of the operators of spins $s_i$ and $s_j$ (obviously $R_{ij} = R_{ji}$) and $Q_{ij}$ counts the contractions of $X$-derivatives of the operators for $s_i$ with the exponent $e^{ipX}$ in the operator for spin $s_j$. Straightforward computation gives:

$$\begin{aligned}
&A_X(p_1,...p_5|w;z;\xi) = \\
&\sum_{\{s_i-1=\sum_j(R_{ij}+Q_{ij})\}} \{\frac{6(s_1-1)!(s_2-1)!(s_3-1)!(\sum s-3)!}{\prod_{i=1}^{4}\prod_{j=2;j>i}^{5} R_{ij}!\prod_{k=1}^{5}\prod_{l=1;k\neq l}^{5} Q_{kl}!} \\
&\times(i)^{\sum_{i,j=1;i\neq j}^{5} Q_{ij}} \\
&\times(-1)^{\sum_{j=1}^{5}(Q_{1j}+Q_{2j})-Q_{21}+Q_{43}+Q_{43}+Q_{45}+Q_{53}} \\
&w^{-\sum_{j=2}^{5}(2R_{1j}+Q_{1j}+Q_{1j})} \\
&\times(1-z)^{p_3p_4-2R_{24}-Q_{24}-Q_{42}}(1-\xi)^{p_2p_4-2R_{25}-Q_{25}-Q_{52}} \\
&\times(z-\xi)^{p_2p_3-2R_{45}-Q_{45}-Q_{54}} z^{p_1p_3-2R_{34}-Q_{34}-Q_{43}} \\
&\xi^{p_1p_2-2R_{35}-Q_{35}-Q_{53}} \\
&\times p_1^{m_{Q_{12}+Q_{14}+Q_{15}+1}}...p_1^{m_{Q_{12}+Q_{14}+Q_{15}+Q_{13}}} \\
&p_1^{n_{Q_{21}+Q_{24}+Q_{25}+1}}...p_1^{n_{Q_{21}+Q_{24}+Q_{25}+Q_{23}}} \\
&p_1^{q_{Q_{41}+Q_{42}+Q_{45}+1}}...p_1^{q_{Q_{41}+Q_{42}+Q_{45}+Q_{43}}} \\
&p_1^{r_{Q_{51}+Q_{52}+Q_{54}+1}}...p_1^{r_{Q_{51}+Q_{52}+Q_{54}+Q_{53}}} \\
&p_2^{m_{Q_{12}+Q_{14}+1}}...p_2^{m_{Q_{12}+Q_{14}+Q_{15}}} p_2^{n_{Q_{21}+Q_{24}+1}}...p_2^{n_{Q_{21}+Q_{24}+Q_{25}}} \\
&p_2^{q_{Q_{41}+Q_{42}+1}}...p_2^{q_{Q_{41}+Q_{42}+Q_{45}}} \\
&p_2^{t_{Q_{31}+Q_{32}+Q_{34}+1}}...p_2^{t_{Q_{31}+Q_{32}+Q_{34}+Q_{35}}} \\
&\times p_3^{m_{Q_{12}+1}}...p_3^{m_{Q_{12}+Q_{14}}} p_3^{n_{Q_{21}+1}}...p_3^{n_{Q_{21}+Q_{24}}} \\
&p_3^{r_{Q_{51}+Q_{52}+1}}...p_3^{r_{Q_{51}+Q_{52}+Q_{54}}} p_3^{t_{Q_{31}+Q_{32}+1}}...p_3^{t_{Q_{31}+Q_{32}+Q_{34}}} \\
&p_4^{m_1}...p_4^{m_{Q_{12}}} p_4^{q_{Q_{41}+1}}...p_4^{q_{Q_{41}+Q_{42}}} p_4^{r_{Q_{51}+1}} \times ...
\end{aligned}$$

$$...\times p_4^{r_{Q_{51}+Q_{52}}} p_4^{t_{Q_{31}+1}}...p_4^{t_{Q_{31}+Q_{32}}}$$
$$p_5^{n_1}...p_5^{n_{Q_{21}}} p_5^{q_1}...p_5^{q_{Q_{41}}} p_5^{r_1}...p_5^{r_{Q_{51}}} p_5^{t_1}...p_5^{t_{Q_{31}}}$$
$$\times\eta^{m_{\Sigma_j Q_{1j}+1}|n_{\Sigma_j Q_{2j}+1}}...\eta^{m_{\Sigma_j Q_{1j}+R_{12}}|n_{\Sigma_j Q_{2j}+R_{12}}}$$
$$\eta^{m_{\Sigma_j Q_{1j}+1+R_{12}}|q_{\Sigma_j Q_{4j}+1}}...\eta^{m_{\Sigma_j Q_{1j}+R_{12}+R_{14}}|q_{\Sigma_j Q_{4j}+R_{14}}}$$
$$\eta^{m_{\Sigma_j Q_{1j}+R_{12}+R_{14}+1}|r_{\Sigma_j Q_{5j}+1}}\times...$$
$$...\times\eta^{m_{\Sigma_j Q_{1j}+R_{12}+R_{14}+R_{15}}|r_{\Sigma_j Q_{5j}+R_{15}}}$$
$$\eta^{m_{\Sigma_j Q_{1j}+R_{12}+R_{14}+R_{15}+1}|t_{\Sigma_j Q_{3j}+1}}\times...$$
$$...\times\eta^{m_{\Sigma_j Q_{1j}+R_{12}+R_{14}+R_{15}+R_{13}}|t_{\Sigma_j Q_{3j}+R_{13}}}$$
$$\eta^{n_{\Sigma_j Q_{2j}+1+R_{12}}|q_{\Sigma_j Q_{4j}+R_{14}+1}}\times...$$
$$...\times\eta^{m_{\Sigma_j Q_{2j}+R_{12}+R_{24}}|q_{\Sigma_j Q_{4j}+R_{14}+R_{24}}}$$
$$\eta^{n_{\Sigma_j Q_{2j}+1+R_{12}+R_{24}}|r_{\Sigma_j Q_{5j}+R_{15}+1}}\times...$$
$$...\times\eta^{m_{\Sigma_j Q_{2j}+R_{12}+R_{24}+R_{25}}|r_{\Sigma_j Q_{5j}+R_{15}+R_{25}}}$$
$$\eta^{n_{\Sigma_j Q_{2j}+1+R_{12}+R_{24}+R_{25}+1}|t_{\Sigma_j Q_{3j}+R_{31}+1}}\times...$$
$$...\times\eta^{m_{\Sigma_j Q_{2j}+R_{12}+R_{24}+R_{25}}|t_{\Sigma_j Q_{3j}+R_{31}+R_{32}}}$$
$$\eta^{q_{\Sigma_j Q_{4j}+1+R_{41}+R_{42}}|r_{\Sigma_j Q_{5j}+R_{51}+R_{52}+1}}\times...$$
$$...\times\eta^{q_{\Sigma_j Q_{4j}+R_{41}+R_{42}+R_{45}}|r_{\Sigma_j Q_{5j}+R_{15}+R_{25}+R_{54}}}$$
$$\eta^{q_{\Sigma_j Q_{4j}+1+R_{41}+R_{42}+R_{45}}|t_{\Sigma_j Q_{3j}+R_{31}+R_{32}+1}}...$$
$$...\times\eta^{q_{\Sigma_j Q_{4j}+R_{41}+R_{42}+R_{45}+R_{43}}|t_{\Sigma_j Q_{3j}+R_{31}+R_{32}+R_{34}}}$$
$$\eta^{r_{\Sigma_j Q_{5j}+1+R_{51}+R_{52}+R_{54}}|t_{\Sigma_j Q_{3j}+R_{31}+R_{32}+R_{34}+1}}...\eta^{r_{\Sigma s-3}|t_{s_3-1}} \quad (4.145)$$

where in our notations $\eta^{m_i|n_j}$ is Minkowski tensor and $\sum_j A_{ij} \equiv \sum_{j=1;j\neq i}^{5} A_{ij}$.

This concludes the $X$-part evaluation. Next, consider the ghost part of the amplitude. The relevant correlator is

$$A_{gh}(w;z;\xi) =< ce^{-s_1\phi}(w)ce^{-s_2\phi}(1)e^{(u-2)\phi}B^{(2u-2)}_{2\phi-2\chi-\sigma}(z)$$
$$e^{(\Sigma s-u)\phi}B^{(2\Sigma s-u)}_{2\phi-2\chi-\sigma}(\xi)ce^{-s_3\phi}(0) > \quad (4.146)$$

To calculate it, introduce again the characteristic partitions:

$$2(u-2) = 4 = \alpha_1+\alpha_2+\alpha_3+\alpha_4^{(1)}+\alpha_4^{(2)}$$
$$2(\Sigma s-u)\equiv 2(\Sigma s-4) = \beta_1+\beta_2+\beta_3+\beta_4^{(1)}+\beta_4^{(2)} \quad (4.147)$$

where the integers $\alpha_j (j = 1,2,3)$ refer to the OPE singularity orders $\sim (z - z_j)^{-\alpha_j}$ due to contractions of $B^{(2u-2)}_{2\phi-2\chi-\sigma}(z)$ with the ghost exponents $ce^{-s_j\phi}(z_j)$; $\beta_j (j = 1,2,3)$ refer to the OPE singularities $\sim (\xi - z_j)^{-\beta_j}$ due to contractions of $B^{(2\Sigma s-u)}_{2\phi-2\chi-\sigma}(\xi)$ with $ce^{-s_j\phi}(z_j)$; $\alpha_4^{(1)}$ $\beta_4^{(1)}$ refer to the OPE singularities of the Bell polynomials at $z$ and $\xi$ with the opposite ghost exponents $e^{(\Sigma s-u)\phi}(\xi)$ and $B^{(2u-2)}_{2\phi-2\chi-\sigma}(z)$ respectively and, finally $\alpha_4^{(2)}$ and $\beta_4^{(2)}$ refer to the OPE due to contractions between Bell polynomials at $z$ and $\xi$, i.e.

$$\begin{aligned} & B^{(2u-2)}_{2\phi-2\chi-\sigma}(z) B^{(2\Sigma s-u)}_{2\phi-2\chi-\sigma}(\xi) \\ & \sim \sum_{\alpha_4^{(2)},\beta_4^{(2)}} (z-\xi)^{-\alpha_4^{(2)}-\beta_4^{(2)}} \lambda(\alpha_4^{(2)},\beta_4^{(2)}) \\ & \times : B^{(2u-2-\alpha_4^{(2)})}_{2\phi-2\chi-\sigma}(z) B^{(2\Sigma s-u-\beta_4^{(2)})}_{2\phi-2\chi-\sigma}(\xi) : \end{aligned} \tag{4.148}$$

where $\lambda(\alpha_4^{(2)},\beta_4^{(2)})$ are the structure constants that can be computed (see the Chapter 5). The correlator (4.148) is thus equal to the universal factor due to contractions between the ghost exponents, multiplied by the sum stemming from contractions of the Bell polynomials with themselves and with various exponents, with each term in the sum corresponding to some of the partitions (4.147) with the correlator being given by the sum over the partitions. Using the explicit expressions for the operator products, needed to compute the correlator, it is straightforward to compute the ghost correlator, with the result given by:

$$\begin{aligned} & A_{gh}(w;z;\xi) \\ & = \sum_{[partitions: 2(u-2)\equiv 4|\alpha_1+\alpha_2+\alpha_3+\alpha_4^{(1)}+\alpha_4^{(2)}]} \\ & \sum_{[partitions: 2(\Sigma s-u)\equiv 2(\Sigma s-4)|\beta_1+\beta_2+\beta_3+\beta_4^{(1)}+\beta_4^{(2)}]} \\ & \prod_{j=1}^{3} \frac{((2s_j-1)!)^2}{\alpha_j!\beta_j!(2s_j-1-\alpha_j)!(2s_j-1-\beta_j)!} \\ & \times \frac{\Gamma(3-2u)\Gamma(1+2u-2\Sigma s)}{\Gamma(3-2u-\alpha_4^{(1)})\Gamma(1+2u-2\Sigma s-\beta_4^{(1)})}\Big|_{u\to 4} \\ & \times (-1)^{\alpha_4^{(2)}+1} w^{s_1^2+2-\alpha_1-\beta_1}(1-z)^{2s_2-\alpha_2}(1-\xi)^{s_2(\Sigma s-4)-\beta_2} \\ & \times z^{2s_3-\alpha_3}\xi^{s_3(\Sigma s-4)-\beta_3}(z-\xi)^{2(4-\Sigma s)-\alpha_4^{(1)}-\alpha_4^{(2)}-\beta_4^{(1)}-\beta_4^{(2)}} \end{aligned} \tag{4.149}$$

This concludes the computation of the ghost factor, contributing to the integrand of the 5-point amplitude. Finally, we are left with computing the $\psi$-factor, given by

$$A_{\psi}(w;z;\xi) = \\ < \prod_{j_1=0}^{s_1-3} \partial^{j_1}\psi_{\alpha_{j_1}}(w) \prod_{j_2=0}^{s_2-3} \partial^{j_2}\psi_{\beta_{j_2}}(1) \prod_{j_3=0}^{u-3} \partial^{j_3}\psi_{\gamma_{j_3}}(z) \\ \prod_{j_4=0}^{\Sigma s-u-1} \partial^{j_4}\psi_{\lambda_{j_4}}(\xi) \prod_{j_5=0}^{s_3-3} \partial^{j_5}\psi_{\sigma_{j_5}}(0) > \tag{4.150}$$

This amplitude would be actually the most tedious one to compute for generic spin values, since there seems to be no way to systemize it in terms of sums over partitions of the spin values, as it has been done for the previous correlators. However, assuming that $u$ is the minimal spin value of the operators, it simplifies significantly for the minimal possible $u = 4$ value (for $u < 4$ it vanishes identically). The simplification is that for $u = 4$, all the $\psi$-fields of 4 operators have to couple to the $\psi$-fields of the operator with the highest spin value, given by $s_{max} = \Sigma s - 2$. Technically, this means that in the corresponding space-time amplitude all the 4 out of 5 $\Omega^{s-1|t}$ extra-field's $t$-indices $\alpha, \beta, \gamma, \sigma$ have to contract to the $\lambda$-index of the highest spin field $\Omega^{s_{max}-1|\lambda}$. The computation performed below is still possible to perform for $u > 4$, e.g. for $u = 5$ or 6, however, for $u > 4$ the locality of the interaction vertex in space-time would be broken by the RG flows from lower orders despite the locality of the scattering amplitude.

In case of $u = 4$, it is convenient to make the following definition. Let $i(\alpha_k)$ describe the contraction of the worldsheet fermion $\partial^k\psi_{\alpha_k}$ of the spin $s_1$ operator to the fermion $\partial^{i(\alpha_k)}\psi_{\lambda_{i(\alpha_k)}}$ of the spin $s_{max}$ operator. Similarly, let $i(\beta_k)$ describe the contractions between the worldsheet fermions of the $s_2$ and $s_{max}$ operators, and so on. Clearly, $i(\alpha) \neq i(\beta) \neq i(\gamma) \neq i(\sigma)$ for all values of $\alpha, \beta, \gamma, \sigma$ and $0 \leq i(\alpha, \beta, \gamma, \sigma) \leq s_{max} - 3$. Then it is natural to express the $\psi$-correlator in terms of the sum over the permutations of different $i$'s , or, equivalently, in terms of sum over all possible length $s_{max} -$ 2 orderings of unequal integer numbers from 0 to $s_{max} - 3$.

The $A_{\psi}$ correlator is then straightforward to compute in terms of such a sum, with the result given by

$$A_\psi(w,z,\xi)=(-1)^{s_1s_2+s_1+s_2+s_3}$$
$$\times\sum_{[permut.:\{,i(\alpha_0)...i(\alpha_{s_1-3}),i(\beta_0)...i(\beta_{s_2-3}),i(\gamma_0),i(\gamma_1),i(\sigma_0)...i(\sigma_{s_3-3})\}]}$$
$$\{(-1)^{\pi(\{,i(\alpha_0)...i(\alpha_{s_1-3}),i(\beta_0)...i(\beta_{s_2-3}),i(\gamma_0),i(\gamma_1),i(\sigma_0)...i(\sigma_{s_3-3})\})}$$
$$\times(1+i(\gamma_0))!(2+i(\gamma_1))!$$
$$\prod_{k_1=1}^{s_1-2}(k_1+i(\alpha_{k_1}-1))!\prod_{k_2=1}^{s_2-2}(k_2+i(\beta_{k_2}-1))!$$
$$\times\prod_{k_3=1}^{s_3-2}(k_3+i(\beta_{k_3}-1))!w^{\frac{1}{2}(s_1-2)^2+\sum_{l_1=0}^{s_1-3}i(\alpha_{l_1})}$$
$$(1-\xi)^{\frac{1}{2}(s_2-2)^2+\sum_{l_2=0}^{s_2-3}i(\beta_{l_2})}\xi^{\frac{1}{2}(s_3-2)^2+\sum_{l_3=0}^{s_2-3}i(\sigma_{l_3})}(z-\xi)^{2+i(\gamma_0)+i(\gamma_1)}\}$$
$$(4.151)$$

where

$$\pi(\{i(\alpha_0)...i(\alpha_{s_1-3}),i(\beta_0)...i(\beta_{s_2-3}),i(\gamma_0),i(\gamma_1),i(\sigma_0)...i(\sigma_{s_3-3})\})$$

is the minimal number of the permutations it takes to convert the ordering

$$\{i(\alpha_0)...i(\alpha_{s_1-3}),i(\beta_0)...i(\beta_{s_2-3}),i(\gamma_0),i(\gamma_1),i(\sigma_0)...i(\sigma_{s_3-3})\}$$

in the argument of $\pi$ into the reference ordering

$$\{0,1,2,.....,s_{max}-3\}$$

of $s_{max}-2$ integers. This concludes the computation of the $\psi$-factor contribuiting to the integrand of the amplitude. The final remaining step to determine the amplitude describing the 5-point higher spin interaction is to perform the double worldsheet integration over the positions of the integrated vertices, using (4.133). Setting the limit $w\to\infty$ and contracting the result with the space-time frame-like higher spin fields (in the integral (4.152) it is convenient to choose the reference $u$-points $u=0$ in the the both of the homotopy transformations for the integrated vertex operators). Combining $A_X$, $A_\psi$ and $A_{gh}$ factors together, evaluating the integral and contracting with the frame-like space-time fields, we obtain the overall quintic amplitude, given by

$$A(p_1,...,p_5)$$
$$=6\pi\Omega_{m_1...m_{s_1-1}|\lambda_0...\lambda_{s_1-3}}(p_5)\Omega_{n_1...n_{s_2-1}|\lambda_{s_1-2}...\lambda_{s_1+s_2-5}}(p_4)$$
$$\Omega_{q_1q_2q_3|\lambda_{s_1+s_2-4}\lambda_{s_1+s_2-3}}(p_3)\Omega^{\lambda_0...\lambda_{\Sigma s-5}}_{r_1...r_{\Sigma s-3}}(p_2)$$
$$\times\Omega_{t_1...t_{s_3-1}|\lambda_{s_1+s_2-2}...\lambda_{\Sigma s-5}}(p_1)$$
$$\times\sum_{partitions:s_i-1|\sum_{j=1,j\neq i}^5(R_{ij}+Q_{ij});i=1,...,5}$$

$$\sum_{partitions:2(u-2)\equiv 4|\alpha_1+\alpha_2+\alpha_3+\alpha_4^{(1)}+\alpha_4^{(2)}}$$
$$\sum_{partitions:2(\Sigma s-u)\equiv 2(\Sigma s-4)|\beta_1+\beta_2+\beta_3+\beta_4^{(1)}+\beta_4^{(2)}}$$
$$\sum_{permutations:\{,i(\alpha_0)...i(\alpha_{s_1-3}),i(\beta_0)...i(\beta_{s_2-3}),i(\gamma_0),i(\gamma_1),i(\sigma_0)...i(\sigma_{s_3-3})\}}$$
$$\{(-1)^{\pi(\{i(\alpha_0)...i(\alpha_{s_1-3}),i(\beta_0)...i(\beta_{s_2-3}),i(\gamma_0),i(\gamma_1),i(\sigma_0)...i(\sigma_{s_3-3})\})}$$
$$\times(-1)^{+s_1s_2+s_1+s_2+s_3+\alpha_4^{(2)}}$$
$$\times(1+i(\gamma_0))!(2+i(\gamma_1))!\prod_{k_1=1}^{s_1-2}(k_1+i(\alpha_{k_1}-1))!$$
$$\prod_{k_2=1}^{s_2-2}(k_2+i(\beta_{k_2}-1))!\prod_{k_3=1}^{s_3-2}(k_3+i(\beta_{k_3}-1))!$$
$$\times\prod_{j=1}^{3}\frac{((2s_j-1)!)^2}{\alpha_j!\beta_j!(2s_j-1-\alpha_j)!(2s_j-1-\beta_j)!}$$
$$\times\frac{\Gamma(3-2u)\Gamma(1+2u-2\Sigma s)}{\Gamma(3-2u-\alpha_4^{(1)})\Gamma(1+2u-2\Sigma s-\beta_4^{(1)})}|_{u\to 4}$$
$$\times\frac{(s_1-1)!(s_2-1)!(s_3-1)!(\sum s-3)!}{\prod_{i=1}^{4}\prod_{j=2;j>i}^{5}R_{ij}!\prod_{k=1}^{5}\prod_{l=1;k\neq l}^{5}Q_{kl}!}$$
$$(i)^{\sum_{i,j=1;i\neq j}^{5}Q_{ij}}(-1)^{\sum_{j=1}^{5}(Q_{1j}+Q_{2j})-Q_{21}+Q_{43}+Q_{43}+Q_{45}+Q_{53}}$$
$$\times p_1^{m_{Q_{12}+Q_{14}+Q_{15}+1}}...p_1^{m_{Q_{12}+Q_{14}+Q_{15}+Q_{13}}}$$
$$\times p_1^{n_{Q_{21}+Q_{24}+Q_{25}+1}}...p_1^{n_{Q_{21}+Q_{24}+Q_{25}+Q_{23}}}$$
$$\times p_1^{q_{Q_{41}+Q_{42}+Q_{45}+1}}...p_1^{q_{Q_{41}+Q_{42}+Q_{45}+Q_{43}}}$$
$$\times p_1^{r_{Q_{51}+Q_{52}+Q_{54}+1}}...p_1^{r_{Q_{51}+Q_{52}+Q_{54}+Q_{53}}}$$
$$\times p_2^{m_{Q_{12}+Q_{14}+1}}...p_2^{m_{Q_{12}+Q_{14}+Q_{15}}}p_2^{n_{Q_{21}+Q_{24}+1}}...p_2^{n_{Q_{21}+Q_{24}+Q_{25}}}$$
$$\times p_2^{q_{Q_{41}+Q_{42}+1}}...p_2^{q_{Q_{41}+Q_{42}+Q_{45}}}p_2^{t_{Q_{31}+Q_{32}+Q_{34}+1}}...p_2^{t_{Q_{31}+Q_{32}+Q_{34}+Q_{35}}}$$
$$\times p_3^{m_{Q_{12}+1}}...p_3^{m_{Q_{12}+Q_{14}}}p_3^{n_{Q_{21}+1}}...p_3^{n_{Q_{21}+Q_{24}}}$$
$$\times p_3^{r_{Q_{51}+Q_{52}+1}}...p_3^{r_{Q_{51}+Q_{52}+Q_{54}}}p_3^{t_{Q_{31}+Q_{32}+1}}...p_3^{t_{Q_{31}+Q_{32}+Q_{34}}}$$
$$\times p_4^{m_1}...p_4^{m_{Q_{12}}}p_4^{q_{Q_{41}+1}}...p_4^{q_{Q_{41}+Q_{42}}}p_4^{r_{Q_{51}+1}}...p_4^{r_{Q_{51}+Q_{52}}}p_4^{t_{Q_{31}+1}}...p_4^{t_{Q_{31}+Q_{32}}}$$
$$\times p_5^{n_1}...p_5^{n_{Q_{21}}}p_5^{q_1}...p_5^{q_{Q_{41}}}p_5^{r_1}...p_5^{r_{Q_{51}}}p_5^{t_1}...p_5^{t_{Q_{31}}}$$

$$\frac{\Gamma[p_2p_3+9-T_{45}-2\Sigma s-\beta_2-\frac{1}{2}(s_2-2)^2-\sum_{k=1}^{s_2-2}i(\beta_{k-1})]}{\Gamma[-p_2p_4+T_{25}-s_2(\Sigma s-4)+\frac{1}{2}(s_2-2)^2+\beta_2+\sum_{k=1}^{s_2-2}i(\beta_{k-1})]}$$

$$\frac{\Gamma[p_3p_4-T_{24}+2s_2-\alpha_2]}{sin[\pi(p_2p_4-T_{25}+s_2(\Sigma s-4)-\frac{1}{2}(s_2-2)^2-\beta_2-\sum_{k=1}^{s_2-2}i(\beta_{k-1}))]}$$

$$\times\Gamma[p_2p_4+p_2p_3+p_1p_3+6-T_{25}-T_{45}-T_{34}+$$

$$(s_2-2)(\Sigma s-4)-\frac{1}{2}(s_2-2)^2+2s_3$$

$$-\alpha_3-\sum_{k=1}^{s_2-2}i(\beta_{k-1})-i(\gamma_0)-i(\gamma_1)]$$

$$\Gamma^{-1}[p_2p_4+p_2p_3+2-T_{25}-T_{45}+(s_2-2)(\Sigma s-4)$$

$$-\frac{1}{2}(s_2-2)^2-\sum_{k=1}^{s_2-2}i(\beta_{k-1})-i(\gamma_0)-i(\gamma_1)]$$

$$\times\Gamma[p_2p_4+p_2p_3+p_1p_3+p_3p_4+6-T_{24}$$

$$+2s_2$$

$$-\alpha_2-T_{25}-T_{45}-T_{34}$$

$$+(s_2-2)(\Sigma s-4)-\frac{1}{2}(s_2-2)^2+2s_3$$

$$-\alpha_3-\sum_{k=1}^{s_2-2}i(\beta_{k-1})-i(\gamma_0)-i(\gamma_1)]$$

$$\times{}_3F_2[-p_1p_2+T_{35}-(s_3+2)(\Sigma s-4)+\frac{1}{2}(s_3-2)^2$$

$$+\beta_3+\sum_{k=1}^{s_3-2}i(\beta_{k-1});$$

$$p_2p_4+1-T_{25}+s_2(\Sigma s-4)-\frac{1}{2}(s_2-2)^2-\beta_2-\sum_{k=1}^{s_2-2}i(\beta_{k-1});$$

$$p_2p_4+p_2p_3+p_1p_3+6-T_{25}-T_{45}-T_{34}+(s_2-2)(\Sigma s-4)$$

$$-\frac{1}{2}(s_2-2)^2+2s_3-\alpha_3-\sum_{k=1}^{s_2-2}i(\beta_{k-1})-i(\gamma_0)-i(\gamma_1);$$

$$p_2p_4+p_2p_3+2-T_{25}-T_{45}+(s_2-2)(\Sigma s-4)$$

$$-\frac{1}{2}(s_2-2)^2-\sum_{k=1}^{s_2-2}i(\beta_{k-1})-i(\gamma_0)-i(\gamma_1);$$

$$
\begin{aligned}
& p_2p_4 + p_2p_3 + p_1p_3 + p_3p_4 + 6 \\
& -T_{25} - T_{45} - T_{34} - T_{24} + (s_2 - 2)(\Sigma s - 4) \\
& -\frac{1}{2}(s_2 - 2)^2 + 2s_2 + 2s_3 - \alpha_2 - \alpha_3 \\
& -\sum_{k=1}^{s_2-2} i(\beta_{k-1}) - i(\gamma_0) - i(\gamma_1); 1] \\
& \times \eta^{m\sum_j Q_{1j}+1|n\sum_j Q_{2j}+1} ... \eta^{m\sum_j Q_{1j}+R_{12}|n\sum_j Q_{2j}+R_{12}} \\
& \eta^{m\sum_j Q_{1j}+1+R_{12}|q\sum_j Q_{4j}+1} ... \eta^{m\sum_j Q_{1j}+R_{12}+R_{14}|q\sum_j Q_{4j}+R_{14}} \\
& \eta^{m\sum_j Q_{1j}+R_{12}+R_{14}+1|r\sum_j Q_{5j}+1} \times ... \\
& ... \times \eta^{m\sum_j Q_{1j}+R_{12}+R_{14}+R_{15}|r\sum_j Q_{5j}+R_{15}} \\
& \eta^{m\sum_j Q_{1j}+R_{12}+R_{14}+R_{15}+1|t\sum_j Q_{3j}+1} \times ... \\
& ... \times \eta^{m\sum_j Q_{1j}+R_{12}+R_{14}+R_{15}+R_{13}|t\sum_j Q_{3j}+R_{13}} \\
& \eta^{n\sum_j Q_{2j}+1+R_{12}|q\sum_j Q_{4j}+R_{14}+1} \times ... \\
& ... \times \eta^{m\sum_j Q_{2j}+R_{12}+R_{24}|q\sum_j Q_{4j}+R_{14}+R_{24}} \\
& \eta^{n\sum_j Q_{2j}+1+R_{12}+R_{24}|r\sum_j Q_{5j}+R_{15}+1} \times ... \\
& ... \times \eta^{m\sum_j Q_{2j}+R_{12}+R_{24}+R_{25}|r\sum_j Q_{5j}+R_{15}+R_{25}} \\
& \eta^{n\sum_j Q_{2j}+1+R_{12}+R_{24}+R_{25}+1|t\sum_j Q_{3j}+R_{31}+1} \times ... \\
& ... \times \eta^{m\sum_j Q_{2j}+R_{12}+R_{24}+R_{25}|t\sum_j Q_{3j}+R_{31}+R_{32}} \\
& \eta^{q\sum_j Q_{4j}+1+R_{41}+R_{42}|r\sum_j Q_{5j}+R_{51}+R_{52}+1} \times ... \\
& ... \times \eta^{q\sum_j Q_{4j}+R_{41}+R_{42}+R_{45}|r\sum_j Q_{5j}+R_{15}+R_{25}+R_{54}} \\
& \eta^{q\sum_j Q_{4j}+1+R_{41}+R_{42}+R_{45}|t\sum_j Q_{3j}+R_{31}+R_{32}+1} \times ... \\
& ... \times \eta^{q\sum_j Q_{4j}+R_{41}+R_{42}+R_{45}+R_{43}|t\sum_j Q_{3j}+R_{31}+R_{32}+R_{34}} \\
& \eta^{r\sum_j Q_{5j}+1+R_{51}+R_{52}+R_{54}|t\sum_j Q_{3j}+R_{31}+R_{32}+R_{34}+1} ... \eta^{r\Sigma s-3|t_{s_3-1}}\}
\end{aligned}
\tag{4.152}
$$

where

$$T_{ij} = 2R_{ij} + Q_{ij} + Q_{ji} \tag{4.153}$$

and the additional constraint is imposed on the partitions:

$$\sum_{j=2}^{5} Q_{j1} + \sum_{k=1}^{s_1-2} i(\alpha_{k-1}) = 3 + \frac{s_1(s_1 - 1)}{2} \tag{4.154}$$

stemming from the fact that only $w^0$-terms conmtribute to the amplitude, with all others vanishing in the limit $w \to \infty$. This concludes the computation of the five-point amplitude for the localized quintic interaction. In the next section we shall discuss the construction of the quintic higher spin vertex, related to this amplitude.

## 4.9 Reading off the Quintic Vertex

In the previous section we have computed the five-point worldsheet amplitude, related to the interaction of masssless higher spin in the quintic order. By itself, this amplitude does not describe yet the 5-point interaction vertex in the low energy effective action: it is only gauge-invariant under gauge (BRST) transformations at the linearized level. To read off the interaction vertex with the full gauge symmetry, one has to subtract from it the terms , produced as a result of the worldsheet RG flows of the effective action's terms at lower orders, such as cubic and quartic. Generally speaking, there are two possible sources of such terms at the quintic order: the flow of the quartic vertex in the leading $\alpha'$ order and the flow of the cubic vertex in the subleading $\alpha'$ order. In general, computation of these flow terms would be quite complicated; in particular they would involve the flows stemming from all the diversity of the quartic vertices which by themselves are tedious and complex. Moreover, since the quartic interactions are generally nonlocal, one would generally expect the flow terms to retain these nonlocalities, destroying the local structure of the amplitude. Fortunately, however, for the spin combinations considered below things again get drastically simplified: as we pointed out above, the four-point correlation functions contributing to the worldsheet $\beta$-function of the space-time frame-like field with the highest spin value, relevant to the flow terms at the quintic order, vanish identically, as it is impossible to accommodate the full contractions of the $\psi$-fermions consistently with the ghost number balance restrictions. Therefore the quartic interactions, relevant to the flow terms at the quintic order, stem themselves from the flows from the previous (cubic) order. For this reason, the flow contributions are reduced to the double composition of the RG flows of the cubic terms which structure is relatively simple to control. As the structure of the cubic higher spin vertices is determined by the structure constants of the higher spin algebra and the quintic vertex is cubic in the structure constants, the effect of the cubic terms flows can be expressed by regularizing the poles in $s_{ij}$-channels present in the amplitude (4.152) due to the Euler's gamma-functions and the hypergeometric function where

$$
\begin{aligned}
&s_{12} = \frac{1}{2}(p_1 - p_2)^2; s_{13} = \frac{1}{2}(p_1 - p_3)^2; s_{14} = \frac{1}{2}(p_1 - p_4)^2 \\
&s_{23} = \frac{1}{2}(p_2 - p_3)^2; s_{24} = \frac{1}{2}(p_2 - p_4)^2; s_{34} = \frac{1}{2}(p_3 + p_4)^2 \\
&\sum_{i,j} s_{ij} = 0
\end{aligned}
\tag{4.155}
$$

are the generalized Mandelstam variables. Subtracting the poles resulting from the flows of the cubic vertices and taking the field theory limit we find the quintic vertex stemming from the $\beta$-function of the highest spin field is given by:

$$
\begin{aligned}
&A(p_1, ..., p_5) = 6\pi \Omega_{m_1...m_{s_1-1}|\lambda_0...\lambda_{s_1-3}}(p_5) \\
&\Omega_{n_1...n_{s_2-1}|\lambda_{s_1-2}...\lambda_{s_1+s_2-5}}(p_4) \\
&\Omega_{q_1 q_2 q_3|\lambda_{s_1+s_2-4}\lambda_{s_1+s_2-3}}(p_3)\Omega^{\lambda_0...\lambda_{\Sigma s-5}}_{r_1...r_{\Sigma s-3}}(p_2) \\
&\Omega_{t_1...t_{s_3-1}|\lambda_{s_1+s_2-2}...\lambda_{\Sigma s-5}}(p_1) \\
&\times \sum_{partitions: s_i - 1|\sum_{j=1, j\neq i}^{5}(R_{ij}+Q_{ij}); i=1,...,5} \\
&\sum_{partitions: 2(u-2)\equiv 4|\alpha_1+\alpha_2+\alpha_3+\alpha_4^{(1)}+\alpha_4^{(2)}} \\
&\sum_{partitions: 2(\Sigma s-u)\equiv 2(\Sigma s-4)|\beta_1+\beta_2+\beta_3+\beta_4^{(1)}+\beta_4^{(2)}} \\
&\sum_{permutations: \{, i(\alpha_0)...i(\alpha_{s_1-3}), i(\beta_0)...i(\beta_{s_2-3}), i(\gamma_0), i(\gamma_1), i(\sigma_0)...i(\sigma_{s_3-3})\}} \\
&\qquad \pi(\{, i(\alpha_0)...i(\alpha_{s_1-3}), i(\beta_0)...i(\beta_{s_2-3}), i(\gamma_0), i(\gamma_1), i(\sigma_0)...i(\sigma_{s_3-3})\}) \\
&\{(-1)^{+s_1 s_2 + s_1 + s_2 + s_3 + \alpha_4^{(2)}} \\
&(1 + i(\gamma_0))!(2 + i(\gamma_1))! \prod_{k_1=1}^{s_1-2}(k_1 + i(\alpha_{k_1} - 1))! \\
&\prod_{k_2=1}^{s_2-2}(k_2 + i(\beta_{k_2} - 1))! \prod_{k_3=1}^{s_3-2}(k_3 + i(\beta_{k_3} - 1))! \\
&\times \prod_{j=1}^{3} \frac{((2s_j - 1)!)^2}{\alpha_j!\beta_j!(2s_j - 1 - \alpha_j)!(2s_j - 1 - \beta_j)!}
\end{aligned}
$$

$$\times\frac{\Gamma(3-2u)\Gamma(1+2u-2\Sigma s)}{\Gamma(3-2u-\alpha_4^{(1)})\Gamma(1+2u-2\Sigma s-\beta_4^{(1)})}|_{u\to 4}$$

$$\times\frac{(s_1-1)!(s_2-1)!(s_3-1)!(\sum s-3)!}{\prod_{i=1}^4\prod_{j=2;j>i}^5 R_{ij}!\prod_{k=1}^5\prod_{l=1;k\neq l}^5 Q_{kl}!}$$

$$(i)^{\sum_{i,j=1;i\neq j}^5 Q_{ij}}(-1)^{\sum_{j=1}^5(Q_{1j}+Q_{2j})-Q_{21}+Q_{43}+Q_{43}+Q_{45}+Q_{53}}$$

$$\times p_1^{m_{Q_{12}+Q_{14}+Q_{15}+1}}...p_1^{m_{Q_{12}+Q_{14}+Q_{15}+Q_{13}}}p_1^{n_{Q_{21}+Q_{24}+Q_{25}+1}}...$$

$$p_1^{n_{Q_{21}+Q_{24}+Q_{25}+Q_{23}}}p_1^{q_{Q_{41}+Q_{42}+Q_{45}+1}}...p_1^{q_{Q_{41}+Q_{42}+Q_{45}+Q_{43}}}$$

$$p_1^{r_{Q_{51}+Q_{52}+Q_{54}+1}}...p_1^{r_{Q_{51}+Q_{52}+Q_{54}+Q_{53}}}$$

$$p_2^{m_{Q_{12}+Q_{14}+1}}...p_2^{m_{Q_{12}+Q_{14}+Q_{15}}}p_2^{n_{Q_{21}+Q_{24}+1}}...p_2^{n_{Q_{21}+Q_{24}+Q_{25}}}$$

$$p_2^{q_{Q_{41}+Q_{42}+1}}...p_2^{q_{Q_{41}+Q_{42}+Q_{45}}}p_2^{t_{Q_{31}+Q_{32}+Q_{34}+1}}...p_2^{t_{Q_{31}+Q_{32}+Q_{34}+Q_{35}}}$$

$$p_3^{m_{Q_{12}+1}}...p_3^{m_{Q_{12}+Q_{14}}}p_3^{n_{Q_{21}+1}}...p_3^{n_{Q_{21}+Q_{24}}}$$

$$p_3^{r_{Q_{51}+Q_{52}+1}}...p_3^{r_{Q_{51}+Q_{52}+Q_{54}}}p_3^{t_{Q_{31}+Q_{32}+1}}...p_3^{t_{Q_{31}+Q_{32}+Q_{34}}}$$

$$\times p_4^{m_1}...p_4^{m_{Q_{12}}}p_4^{q_{Q_{41}+1}}...p_4^{q_{Q_{41}+Q_{42}}}p_4^{r_{Q_{51}+1}}...p_4^{r_{Q_{51}+Q_{52}}}$$

$$p_4^{t_{Q_{31}+1}}...p_4^{t_{Q_{31}+Q_{32}}}p_5^{n_1}...p_5^{n_{Q_{21}}}p_5^{q_1}...p_5^{q_{Q_{41}}}p_5^{r_1}...p_5^{r_{Q_{51}}}p_5^{t_1}...p_5^{t_{Q_{31}}}$$

$$\times[(-9+T_{45}+2\Sigma s+\beta_2+\frac{1}{2}(s_2-2)^2+\sum_{k=1}^{s_2-2}i(\beta_{k-1}))!]^{-1}$$

$$\times L(-9+T_{45}+2\Sigma s+\beta_2+\frac{1}{2}(s_2-2)^2+\sum_{k=1}^{s_2-2}i(\beta_{k-1}))$$

$$\times(-T_{25}+s_2(\Sigma s-4)-\frac{1}{2}(s_2-2)^2-\beta_2$$

$$-\sum_{k=1}^{s_2-2}i(\beta_{k-1}))!(1-T_{24}+2s_2-\alpha_2)!$$

$$\times[5-T_{25}-T_{45}-T_{34}+(s_2-2)(\Sigma s-4)-\frac{1}{2}(s_2-2)^2$$

$$+2s_3-\alpha_3-\sum_{k=1}^{s_2-2}i(\beta_{k-1})-i(\gamma_0)-i(\gamma_1)]!$$

$$[(1-T_{25}-T_{45}+(s_2-2)(\Sigma s-4)-\frac{1}{2}(s_2-2)^2$$

$$-\sum_{k=1}^{s_2-2}i(\beta_{k-1})-i(\gamma_0)-i(\gamma_1))!]^{-1}$$

$$\times[5-T_{24}+2s_2-\alpha_2-T_{25}-T_{45}-T_{34}+(s_2-2)(\Sigma s-4)$$

$$-\frac{1}{2}(s_2-2)^2+2s_3-\alpha_3-\sum_{k=1}^{s_2-2}i(\beta_{k-1})-i(\gamma_0)-i(\gamma_1)]!$$

$$\times {}_3F_2[T_{35} - (s_3+2)(\Sigma s - 4) + \frac{1}{2}(s_3-2)^2 + \beta_3 + \sum_{k=1}^{s_3-2} i(\beta_{k-1});$$

$$1 - T_{25} + s_2(\Sigma s - 4) - \frac{1}{2}(s_2-2)^2 - \beta_2 - \sum_{k=1}^{s_2-2} i(\beta_{k-1});$$

$$6 - T_{25} - T_{45} - T_{34} + (s_2-2)(\Sigma s - 4) - \frac{1}{2}(s_2-2)^2$$

$$+2s_3 - \alpha_3 - \sum_{k=1}^{s_2-2} i(\beta_{k-1}) - i(\gamma_0) - i(\gamma_1);$$

$$2 - T_{25} - T_{45} + (s_2-2)(\Sigma s - 4) - \frac{1}{2}(s_2-2)^2$$

$$- \sum_{k=1}^{s_2-2} i(\beta_{k-1}) - i(\gamma_0) - i(\gamma_1);$$

$$6 - T_{25} - T_{45} - T_{34} - T_{24} + (s_2-2)(\Sigma s - 4)$$

$$-\frac{1}{2}(s_2-2)^2 + 2s_2 + 2s_3$$

$$-\alpha_2 - \alpha_3 - \sum_{k=1}^{s_2-2} i(\beta_{k-1}) - i(\gamma_0) - i(\gamma_1); 1]$$

$$\times \eta^{m\sum_j Q_{1j}+1|n\sum_j Q_{2j}+1} \ldots \eta^{m\sum_j Q_{1j}+R_{12}|n\sum_j Q_{2j}+R_{12}}$$

$$\eta^{m\sum_j Q_{1j}+1+R_{12}|q\sum_j Q_{4j}+1} \ldots \eta^{m\sum_j Q_{1j}+R_{12}+R_{14}|q\sum_j Q_{4j}+R_{14}}$$

$$\eta^{m\sum_j Q_{1j}+R_{12}+R_{14}+1|r\sum_j Q_{5j}+1} \times \ldots$$

$$\ldots \times \eta^{m\sum_j Q_{1j}+R_{12}+R_{14}+R_{15}|r\sum_j Q_{5j}+R_{15}}$$

$$\eta^{m\sum_j Q_{1j}+R_{12}+R_{14}+R_{15}+1|t\sum_j Q_{3j}+1} \times \ldots$$

$$\ldots \times \eta^{m\sum_j Q_{1j}+R_{12}+R_{14}+R_{15}+R_{13}|t\sum_j Q_{3j}+R_{13}}$$

$$\eta^{n\sum_j Q_{2j}+1+R_{12}|q\sum_j Q_{4j}+R_{14}+1} \ldots \eta^{m\sum_j Q_{2j}+R_{12}+R_{24}|q\sum_j Q_{4j}+R_{14}+R_{24}}$$

$$\eta^{n\sum_j Q_{2j}+1+R_{12}+R_{24}|r\sum_j Q_{5j}+R_{15}+1} \times \ldots$$

$$\ldots \times \eta^{m\sum_j Q_{2j}+R_{12}+R_{24}+R_{25}|r\sum_j Q_{5j}+R_{15}+R_{25}}$$

$$\eta^{n\sum_j Q_{2j}+1+R_{12}+R_{24}+R_{25}+1|t\sum_j Q_{3j}+R_{31}+1} \times \ldots$$

$$\ldots \times \eta^{m\sum_j Q_{2j}+R_{12}+R_{24}+R_{25}|t\sum_j Q_{3j}+R_{31}+R_{32}}$$

$$\eta^{q\sum_j Q_{4j}+1+R_{41}+R_{42}|r\sum_j Q_{5j}+R_{51}+R_{52}+1} \times \ldots$$

$$\ldots \times \eta^{q\sum_j Q_{4j}+R_{41}+R_{42}+R_{45}|r\sum_j Q_{5j}+R_{15}+R_{25}+R_{54}}$$

$$
\begin{aligned}
&\eta^{q\sum_j Q_{4j}+1+R_{41}+R_{42}+R_{45}|t\sum_j Q_{3j}+R_{31}+R_{32}+1} \times \ldots \\
&\ldots \times \eta^{q\sum_j Q_{4j}+R_{41}+R_{42}+R_{45}+R_{43}|t\sum_j Q_{3j}+R_{31}+R_{32}+R_{34}} \\
&\eta^{r\sum_j Q_{5j}+1+R_{51}+R_{52}+R_{54}|t\sum_j Q_{3j}+R_{31}+R_{32}+R_{34}+1} \ldots \eta^{r\sum s-3|t_{s_3}-1}\}
\end{aligned}
\tag{4.156}
$$

where

$$
L(n) = \sum_{m=1}^{n} \frac{1}{m} \tag{4.157}
$$

This concludes the evaluation of the quintic interaction vertex and constitutes a very special limit of higher spin quintic interaction, limited to the case when the sum of three spins participating in the interaction roughly equals the sum of the remaining two, or which one must be small anough and another must be large. In this case, the ghost structure of the operators drastically simplifies the calculations and the absence of RG flows from the quartic order, contributing at the fifth order, makes the construction of the interaction 5-vertex from the scattering amplitude a relatively straightforward procedure. Despite the specific choice of the spin values considered in this work, it is remarkable that the structure of the 5-vertex can be extracted from string theory in the low energy limit. One particularly important result is the locality of the quintic amplitude in this limit. This is the intriguing novelty of the quintic interaction all the known examples of the vertices at the previous (quartic) order are essentially nonlocal, and the vertex constructed has no quartic analogue (which vanishes by the ghost/$\psi$-number constraints). This may bear important important implications for the higher spin holography, as the nonlocality of higher spin vertices versus their local counterparts in the dual CFT's is the well-known puzzle.

Our hope is that the string theory may provide efficient and powerful tools to address these issues, as well as to approach the higher orders of higher spin interactions, at least for specific spin values. It is also of interest to find the holographic interpretation of the localization effect, described above. For that, it is on the other hand necessary to have better understanding of the higher-spin nonlocalities from the string theory side. So far, string theory has been able to account for very limited and simplistic types of higher spin nonlocalities only. For that, the class of $V_{s-1|s-3}$-operators may not be sufficient and one may need to find explicit solutions of the operator equations for the generalized zero torsion constraints. These constraints are generally hard to work with in the formalism of the on-shell (first-quantized) string theory. Alternatively, in the string field theory

(SFT) approach the nonlocalities may be naturally encripted in the structure of operators in the cohomologies of the BRST charge shifted by the appropriate analytic solutions in open string field theory. The obvious advantage of this approach is the background independence, which in theory may allow us to penetrate beyond the realm of standard string perturbation theory. Although at this time our understanding of how the analytic solutions work to describe higher spin interaction is still very preliminary and limited, in the end it seems plausible that the language of shifted BRST cohomologies in open string field theory may be the most natural and efficient to understand the nonlocalities in higher spin interactions.

## Chapter 5

# Bosonic Strings, Background Independence and Analytic Solutions

### 5.1 Background Independence - Some Preliminaries

As we noted above, the background independence of string field theory (SFT) is the key property that, in principle, opens the way to string dynamics in space-time geometries other than the flat one, making it a natural instrument to explore the string-inspired holography. The analytic solutions of string field theory equations of motion play the crucial role in this approach. Below, we shall remind the reader the main ideas behind the SFT construction. For the time being, we shall restrict ourselves to open bosonic strings. Let $X^m(\xi_1, \xi_2)$ $(m = 0, ..., D-1)$ be the space-time coordinates where the bosonic string propagates, with $\xi_1, \xi_2$ parametrizing the worldsheet. We start with the first-quantized theory in flat target space, with the action given by

$$S = -\frac{1}{8\pi}\int d^2\xi\sqrt{-\gamma}\gamma^{ab}\partial_a X_m \partial X^m \tag{5.1}$$

where $a, b = 1, 2$ are the worldsheet indices and $\gamma^{ab}$ is the induced metric. Here and elsewhere we use the units $\alpha' = 1$. This action is invariant under reparametrizations and (classically) Weyl transformations. Upon fixing the conformal gauge ($\gamma_a^b = \delta_a^b$) the full matter+ ghost action (plus the Liouville anomaly for noncritical space-time dimensions) the action is given by that of (3.2), where $\mu_0$ is two-dimensional cosmological constant. The Liouville background charge $q$, the constant $B$ and the space-time dimension $D$ are related according to

$$q = B + \frac{1}{B} = \sqrt{\frac{25-D}{3}} \tag{5.2}$$

in order to ensure that the total matter+ghost+Liouville central charge of the system is zero:

$$c = c_{matter} + c_{b-c} + c_{Liouville} = 0 \tag{5.3}$$

The gauge-fixed action (3.2) is invariant under the BRST symmetry transformations:

$$\begin{aligned}\delta_{BRST}X^m &= -c\partial X^m\\ \delta_{BRST}c &= -\partial cc\\ \delta_{BRST}b &= T\end{aligned} \tag{5.4}$$

(with the similar anti-holomorphic transformations) where $T$ is the holomorphic part of the full stress-energy tensor of the system. Similarly for the antiholomorphic part $\bar{T}$. The BRST symmetry transformations are generated by the holomorphic and antiholomorphic BRST charges $Q, \bar{Q}$, such that $Q^2 = \bar{Q}^2 = 0$ with $Q$ given by (1.9) and similarly for $\bar{Q}$. The physical spectrum of open string theory is defined by BRST cohomology, consisting of vertex operators, which are the solutions of the equation:

$$[Q, V] = 0 \tag{5.5}$$

such that

$$V \neq \{Q, U\} \tag{5.6}$$

or

$$\{Q, V\} = 0 \tag{5.7}$$

such that

$$V \neq [Q, U] \tag{5.8}$$

(in closed string theory, the similar conditions with $\bar{Q}$ are imposed). The first equation is the BRST-invariance constraint, the second one is the BRST-nontriviality constraint.

The first class of the solutions referred to as integrated vertex operators and the second unintegrated vertex operators. In open string theory, the solutions of the first class:

$$V = \oint dz W(z) \tag{5.9}$$

turn out to be the primary matter fields $W(z)$ of dimension 1, integrated over the worldsheet boundary (accordingly, for closed strings, the solutions are the primaries of dimension $(1, 1)$ integrated over the bulk of the worldsheet). The solutions of the second class are the matter primary fields of dimension 1 (or (1,1) for closed strings) multiplied by the $c$-ghost field (by $c\bar{c}$ for closed strings). Indeed, it is straightforward to see that any BRST-invariant local (unintegrated) operator of dimension other than 1 is BRST-trivial: using the identity $\{Q, b\} = T$ and the operator product expansion for primary fields of dimension $h$:

$$T(z)\Phi_h(w) = \frac{h\Phi_h(w)}{(z-w)^2} + \frac{\partial\Phi_h(w)}{z-w} + O(z-w)^0 \tag{5.10}$$

and expanding

$$\begin{aligned} T(z) &= \sum_{n=-\infty}^{\infty} \frac{L_n}{z^{n+2}} \\ b(z) &= \sum_{n=-\infty}^{\infty} \frac{b_n}{z^{n+2}} \end{aligned} \tag{5.11}$$

(where $L_n$ are the Virasoro modes) it is easy to check that for any BRST-invariant $\Phi_h$ or $\oint \frac{dz}{2i\pi}\Phi_h$

$$\begin{aligned} \Phi_h &= \frac{1}{h}[Q, b_0\Phi_h] \\ \oint dz\Phi_h &= \frac{1}{h-1}\{b_0, \oint dz\Phi_h\} \end{aligned} \tag{5.12}$$

On the other hand,for $W(z)$ having conformal dimension 1, the operators $cW$ or $\oint \frac{dz}{2i\pi}W$ are only invariant if $W(z)$ is primary in order to commute with $Q$, i.e. their operator products with $T(z)$ do not contain terms more singular than quadratic poles. This altogether is sufficient to establish the form of the general solution to the equations (5.5)-(5.8). There are only two types of operators (that can be constructed out of $X^m(z)$) forming eigenstates of $L_0$, i.e. having definite conformal dimensions: derivatives and exponents of $X$. For this reason, the general solutions for BRST cohomology have the form

$$V = A(p)cP_N(\partial X, ..., \partial^N X)e^{ipX}p^2 = 2(1-N) \tag{5.13}$$

for unintegrated vertex operators and

$$V = A(p)\oint \frac{dz}{2i\pi}P_N(\partial X, ..., \partial^N X)e^{ipX}p^2 = 2(1-N) \tag{5.14}$$

for integrated vertex operators. Here $P_N$ are the polynomials in derivatives of $X$ with each term having conformal dimension $N$ and $A(p)$ are the space-time fields, sourced by the vertex operators (for the sake of brevity, we have suppressed the space-time indices here; for closed strings, these operators have to be multiplied by antiholomorphic part). The solutions (5.14) describe oscillation modes of a single string. Deforming the string action in flat background with the vertex operators (5.14) and imposing the worldsheet conformal invariance constraints on the deformed action leads to space-time constraints on $A(p)$ constituting the low-energy equations of motion on $A$ (i.e. Einstein gravity equations for a graviton or Born-Infeld equations for a massless vector boson). The next step is to introduce the interactions between multiple strings. For this, it is most natural to use the second-quantized formalism. Introduce the string field $\Psi$ defined as formal series in local operators given by all possible combinations of $X$, $b-c$ ghosts and their derivatives:

$$\Psi = \sum_I V_I(X, \partial X, ..., b, c, \partial b, \partial c, ...) \tag{5.15}$$

Then the BRST-cohomology conditions (5.5)-(5.8) can be viewed as analytic solutions of the equation of motion $Q\Psi = 0$ that formally follows from the quadratic action

$$S = \frac{1}{2}\int Q\Psi \star \Psi \equiv << Q\Psi, \Psi >> \equiv < Q\Psi(0) f^{(2)} \circ \Psi(0) > \tag{5.16}$$

where $f^{(2)}(z) = -\frac{1}{z}$ is conformal transformation acting on all the components of $\Psi$ and $< ... >$ denotes standard worldsheet correlators. The solutions of the linear equation $Q\Psi = 0$ are obviously invariant under the shift $\Psi \to \Psi + Q\Lambda$.

The simplest way to introduce interactions between open strings is to deform the string action with the cubic term:

$$\begin{aligned}
S &= \int \{\frac{1}{2}\Psi Q \Psi + \frac{1}{3}\Psi^3\} \\
&\equiv \frac{1}{2} << Q\Psi, \Psi >> + \frac{1}{3} << \Psi, \Psi \star \Psi >> \\
&\equiv \frac{1}{2} < Q\Psi(0) f^{(2)} \circ \Psi(0) > + \\
&\frac{1}{3} < f_1^{(3)} \circ \Psi(0) f_2^{(3)} \circ \Psi(0) f_3^{(3)} \circ \Psi(0) >
\end{aligned} \tag{5.17}$$

where the conformal transformations $f_k^{(3)}$ $(k = 1, 2, 3)$ acting on string fields are compositions of two transformations:

$$\begin{aligned}
f_k^{(3)}(z) &= h(z) \circ g_k^{(3)}(z) \\
g_k^{(3)}(z) &= e^{\frac{2i\pi(k-1)}{3}} (\frac{1-iz}{1+iz})^{\frac{2}{3}} \\
h(z) &= \frac{i(1-z)}{1+z}
\end{aligned} \tag{5.18}$$

The star product of string fields in the cubic interaction term is defined by conformal transformations (5.18) acting on the string fields and describe interaction of open strings in the second-quantized formalism, as follows:

1) The conformal transformations $f_k^{(3)}(z)$ map the worldsheets of 3 separate strings (each with the topology of a half-plane) to 3 wedges of a single unit disc, gluing cyclically the right half of the first string and the left half of the second, the right half of the second and the left of the third, the right of the third and the left of the first, with the midpoints of the strings coincident.

2) The $h(z)$ conformal transformation then maps the disc to the single half-plane with three punctures at $-\sqrt{3}, 0$ and $\sqrt{3}$ which are the insertions for the 3-point worldsheet correlator.

The gluing describes the elementary 3-vertex of interacting strings, corresponding to 2 open strings merging to form the third. The cubic action (5.17) is invariant with respect to the gauge symmetry transformations:

$$\Psi \to Q\Lambda + \Lambda \star \Psi - \Psi \star \Lambda \tag{5.19}$$

The cubic action (5.17) is the action for Witten's open string field theory (OSFT) [72]. It obviously leads to the equation of motion:

$$Q\Psi + \Psi \star \Psi = 0 \tag{5.20}$$

The crucial property of the string field theory, immediately following from this equation, is the background independence. That is, suppose that a string field $\Psi_0$ is the solution of (5.20). Then, the form of the equation (5.20) is invariant under the shift of $\Psi$ by this solution:

$$\Psi \to \tilde{\Psi} = \Psi + \Psi_0 \tag{5.21}$$

provided that the BRST charge is shifted according to

$$Q \to \tilde{Q} = Q + [\Psi_0 \star,] \tag{5.22}$$

i.e. $\tilde{Q}\Psi = Q\Psi + \Psi_0 \star \Psi - \Psi \star \Psi_0$ for any $\Psi$. It is straightforward to check that the new BRST charge $\tilde{Q}$ is nilpotent if $Q$ is nilpotent and $\Psi_0$ is the solution of (5.20):

$$\begin{aligned} &\tilde{Q}\tilde{\Psi} + \tilde{\Psi} \star \tilde{\Psi} = 0 \\ &\tilde{Q}^2 = 0 \end{aligned} \tag{5.23}$$

The cohomology of the new BRST charge $\tilde{Q}$ defines open string theory in a new background, different from the original one. So for example, one can start from string theory in flat space-time, defined by the BRST charge

(1.9). Suppose we are lucky to find the solution $\Psi_0$, defining new nilpotent $\tilde{Q}$, describing strings propagating in new geometrical (curved) background, such as AdS. To identify the new background, one has to determine the cohomology of $\tilde{Q}$ and the resulting constraints on the space-time fields preserving the conformal invariance on the worldsheet (low-energy equations of motion). If the solution $\Psi_0$ describes strings in new space-time geometry, this geometry will manifest itself in the the low-energy effective action (e.g. the vanishing $\beta$-function constraint for a photon will lead to equations of motion in a curved geometry, making it possible to read off the resulting geometry). This makes it possible to explore string dynamics by technically using the original operator algebra of conformal field theory of string theory in flat space. Although looking as a promising program, this approach faces several challenges. First of all, finding analytic solutions to the equation (5.20) is a notoriously difficult problem, due to complexity of the star product. The star product involves conformal transformations of the type (5.18) which are outside $SL(2,R)$ subalgebra and generically act on composite operators that are highly non-primary and require internal normal ordering, resulting in the appearance of objects like higher-derivative Schwarzians in their transformation laws. Second, the structure of the analytic solutions, relevant to non-perturbative string dynamics (e.g. Schnabl's solution for the tachyon vacuum) is usually quite cumbersome, making it hard to work with the cohomology of the shifted BRST charge $\tilde{Q}$. Until recently,a relatively very few examples of analytic OSFT solutions of physical significance were known. One of the first highly nontrivial examples of such solutions have been first found by Martin Schnabl [52–54], with the $b-c$ ghost number 1, which explicit form is

$$\begin{aligned}\Psi_0 &= \sum_{n=0}^{\infty} \sum_{p=-1,2,3,5,7,\ldots} \frac{\pi^p}{2^{n+2p+1}n!}(-1)^n \Lambda_{n+p+1}(D_0 + D_0^{\dagger})^n \tilde{c}_{-p} \\ &+ \sum_{n=0}^{\infty} \sum_{p,q=-1;p+q=odd} \frac{\pi^{p+q}}{2^{n+2p+1}n!}(-1)^{n+q} \\ &\times \Lambda_{n+p+q+2}(B_0 + B_o^{\dagger})(D_0 + D_0^{\dagger})^n \tilde{c}_{-p}\tilde{c}_{-q}\end{aligned} \tag{5.24}$$

where

$$D_n = \oint \frac{dz}{2i\pi}(1+z^2)(arctan(z))^{n+1}T(z) \tag{5.25}$$

are the conformal mappings of the Virasoro generators from the upper half-plane to semi-infinite cylinder of circumference $\pi$ under $f(z) = arctan(z)$:

these operators define conservation laws for sliver [55] their hermitian conjugates are

$$D_n^\dagger = \oint \frac{dz}{2i\pi}(1+z^2)(arccot(z))^{n+1}T(z) \tag{5.26}$$

$$\begin{aligned} B_0 &= \oint \frac{dz}{2i\pi}(1+z^2)(arctan(z))b(z) = b_0 + 2\sum_{k=1}^{\infty}\frac{(-1)^{k+1}}{4k^2-1}b_{2k} \\ B_0^\dagger &= \oint \frac{dz}{2i\pi}(1+z^2)(arccot(z))b(z) \end{aligned} \tag{5.27}$$

is the mapping (and its conjugate) of the $b$-ghost zero-mode to the cylinder. $\Lambda_n$ are the Bernoulli numbers defined according to

$$\frac{x}{e^x - 1} = \sum_{n=0}^{\infty}\frac{\Lambda_n x^n}{n!} \tag{5.28}$$

The explicit expression for these numbers is given in terms of series in partial Bell polynomials:

$$\begin{aligned} \Lambda_n &= n!\sum_{p=1}^{n}(-1)^p B_{n|p}(a_1, ...a_{n-p+1}) \\ a_q &= \frac{1}{q+1}; q = 1, ...n \end{aligned} \tag{5.29}$$

and they obey a number of remarkable identities, such as the Euler identity:

$$(n+1)\Lambda_n = -\sum_{k=2}^{n-2}\frac{n!\Lambda_k\Lambda_{n-k}}{k!(n-k)!} \tag{5.30}$$

and

$$(n-1)\Lambda_n = -\sum_{0\leq p,q\leq n, p+q\leq n}^{n-2}\frac{n!\Lambda_p\Lambda_q}{p!q!(n-p-q)!} \tag{5.31}$$

along with other quadratic, cubic, quartic and higher-order identities. Some of the identities have been observed for the first time by Schnabl, as he studied the analytic solution (5.25) for the tachyonic vacuum. This is not incidental. In fact, all these identities follow from the underlying identities involving Bell polynomials and their productucts. In turn, the Bell polynomial identities can be understood as a number-theoretic pillar for many other classes of analytic solutions existing in string field theory, generalizing Schnabl's solution for the nonperturbative tachyonic vacuum. Relations between Bell polynomials are closely connected, at the same time, to identities between summations over weighted partitions, defining structures of higher-spin amplitudes and interactions. In the next section, we will discuss this connection between analytic solutions in string field theory and higher-spin dynamics.

## 5.2 Analytic String Field Theory Solution for $AdS_3$ Vacuum

One lesson that we learn the Schnabl's analytic solutions is that the structure of the minimum of the nonperturbative tachyon potential can be accessed and studied without knowing the particulars of this potential (in string perturbation theory, that would require calculations in string perturbation theory up to arbitrary order, which is technically unrealistic). The natural question is whether this construction can be extended beyond the scalar field, in particular, the higher spin fields that we focus on. While we know little about higher-spin interactions at higher orders (except for the fact that they must be extremely cumbersome and nonlocal), the question is: can string field theory tell us anything about the ground state of a theory, defined by such interactions, up to an arbitrary order?

In this section, we shall will try to address this question and discuss another nontrivial class of the analytic SFT solutions, providing a model for such a higher-spin vacuum.

It is well-known that the equations of motion of Witten's cubic string field theory (5.20) resemble the Vasiliev's equations in the unfolding formalism in higher-spin theories [76]

$$dW + W \wedge \star W = 0 \tag{5.32}$$

(flatness condition for connection in infinite-dimensional higher-spin algebras) that determine the interactions of the higher-spin gauge fields in this formalism, along with equations for other master fields, containing higher-spin Weyl tensors and auxiliary fields. Higher spin holography strongly hints, however, that this resemblance may be much more than just a formal similarity. The generalized 1-form $W$ (5.32) contains all the higher-spin gauge fields components in $AdS$ spaces which, by holography principle, are related to various multi-index composite operators in the dual CFT's. Any of these CFT's, in turn, must be a low-energy limit of string theory in $AdS_{d+1}$, with the $CFT_d$ correlators reproduced by the worldsheet correlation functions of the vertex operators in $AdS$ string theory, with the space-time fields polarized along the boundary of the $AdS$ space. On the other hand, the second-quantized string field $\Psi$, satisfying the equation (5.20) is nothing but the expansion containing infinite number of modes determined by these vertex operators. Both string fields and higher spin gauge fields in the equations (5.20) and (5.32) are known to be complicated objects to work with. Despite the fact that the higher spin theories in *AdS* spaces can circumvent the restrictions imposed by the Coleman-Mandula's theorem, describing the gauge-invariant higher-spin interactions is a highly nontrivial problem since the gauge symmetry in these theories must be sufficiently powerful in order to eliminate unphysical degrees of freedom. The restrictions imposed by such a gauge symmetry make the construction of the interaction vertices in higher-spin theories a notoriously complicated problem. While there was some progress in classification of the higher-spin 3-vertices over recent years, the structure of the higher-order interactions (such as quartic interactions, presumably related to conformal blocks in dual CFT's) still remains obscure. The structure of these interactions is, however, crucial for our understanding of higher-spin extensions of the holography principle and non-supersymmetric formulation of $AdS/CFT$.

At the same time, the string field theory still remains our best hope to advance towards background-independent formulation of string dynamics. This, in turn, holds the keys to understanding string theories in curved backgrounds, such as AdS. Such string theories are also crucially relevant to holography and gauge-string correspondence, however, little is known about them beyond the semiclassical limit.

Analytic solutions in string field theory appear to be one of the most crucial ingredients in order to approach such string theories in the SFT formalism, using the concept of background independence. To illustrate this, suppose a string field $\Psi_0$ is a solution of the equation (5.17). Then the form of (5.17) is invariant under the shift

$$\Psi \to \tilde{\Psi} = \Psi + \Psi_0 \tag{5.33}$$

with the simultaneous shift of the BRST charge $Q \to \tilde{Q}$ , so that $Q^2 = \tilde{Q}^2 = 0$ and the new nilpotent charge $\tilde{Q}$ defined according to

$$\tilde{Q}\Psi = Q\Psi + \Psi_0 \star \Psi + \Psi \star \Psi_0 \tag{5.34}$$

for any $\Psi$. Then the new BRST charge $\tilde{Q}$ defines the new cohomology, different from that of the original charge $Q$, corresponding to string theory in a new background, depending on the structure of $\Psi_0$. The advantage of this approach is that, in principle, it allows to explore the string theory in new geometrical backgrounds (e.g. in a curved geometry, such as AdS) while technically using the operator products of the old string theory (say, in originally flat background) for the vertex operatos in the new BRST cohomology, defined by $\tilde{Q}$. This formalism is potentially more powerful than the first-quantized formalism, which is background-dependent and where the vertex operator description is essentially limited to the flat space-time and semiclassical limit of curved backgrounds. Unfortunately, however, the major obstacle is that identifying analytic solutions of the equation (5.20) is hard because of the complexity of the star product in (5.16)-(5.18). For this reason, there are not many known examples of analytic solutions having a clear physical interpretation. As was discussed above, one of the most fascinating and well-known solutions, describing the nonperturbative tachyonic vacuum in string theory is of course the class of the Schnabl's solutions [94], later generalized in a number of important papers, in particular, such as [91–93, 95, 96] which were discovered several years ago and in particular used to prove the Sen's conjecture [49,50]. Since that remarkable paper by Schnable [94] there were many other interesting works describing the related SFT solutions, both in cubic theory and in Berkovits SFT theory [91–93, 98] such as algebraic SFT solutions, the analytic solutions describing various nonperturbative processes such as D-brane translations. Despite that, classes of the SFT solutions, relevant to particular geometric backgrounds in string theory, in particular those that would allow us to advance towards consistent formulation of string theories in different space-time geometries, still mostly remain beyond our reach. One reason for this is that the star product in the equation (5.20) is hard to work with in practice. In general, this product is quite different from the conventional Moyal product or the product in the Vasiliev's equations (5.32), although for certain restricted classes of string fields the star product of can be mapped to the Moyal product [104–106]. In general, however, the star product involves the conformal transformations (5.18) that map the string fields living on separate worldsheets to $N$ wedges of a single disc. The behavior of generic string fields (containing all sorts of off-shell non-primary operators) under such global conformal transformations easily wobbles out of control beyond any low-level truncation, making it hard to evaluate the star product by straightforward computation of the correlators in OSFT. There are very few known exceptions to that, such as the wedge states or the special degenerate case of $\Psi$ constrained to primaries and their derivatives However, such fields form too small a subset in the space of all the operators. The known SFT solutions constrained to this subset do exist. However, with the exception of the Schnabl-related class of solutions, they are typically irrelevant to non-perturbative background deformations. At the same time, there exists a sufficiently large class of the operators (far larger than the class of the primary fields) which behaves in a rather compact and controllable way under the conformal transformations (5.18), forming a closed subset of operators under the global conformal transformations. Typically, these operators have the form

$$T^{(N)} = \sum_{k=1}^{N} \sum_{N|N_1...N_k} \lambda^{(N)}_{N_1...N_k} B^{(N_1)}...B^{(N_k)} \tag{5.35}$$

with the sum taken over the partitions of total conformal dimension $N$ of $T^{(N)}$ and with $B^{(N_i)}(\partial X, \partial^2 X, ..., \partial^{N_j} X)$ being the Bell polynomials of rank $N_j$ in the worldsheet derivatives of string or superstring space-time coordinates or the ghost fields. The structure of the correlators of the operators of the form (5.38), as well as their transformation properties under (5.18), with the Bell polynomials satisfying number-theoretical identities generalizing those satisfied by Bernoulli numbers, makes them natural candidates to test for the analytic solutions of (5.20). At the same time, it turns out that the structure constants of higher spin algebras in *AdS* can be realized in terms of the OPE structure constants of the operators of the type (5.36). This makes a natural guess that the SFT solutions of the form (5.36) describe backgrounds with nonperturbative higher-spin configurations stemming from full interacting (to all orders) higher-spin theory in *AdS*. More precisely, this means the following. Suppose that somehow we manage to take a glimpse into full consistently interacting higher-spin theory and the higher spin interactions to all orders. Of course the Lagrangian of such a theory would be immensely complex, with all due restrictions imposed by the gauge invariance, with nonlocalities etc. One would also expect issues with unitarity as well, at least in backgrounds other than AdS. Assume, however, that we managed to identify such a higher-spin action and to solve the equations of motion, i.e. to find the higher-spin configuration minimizing this action. From the string theory point of view, such a background would correspond to a certain conformal fixed point, with vanishing $\beta$-functions of higher-spin vertex operators. An attempt to compute such $\beta$-functions straightforwardly would be hopeless, since that would require summing up contributions from all orders of the string perturbation theory. However, instead of computing the $\beta$-function, one can try to find an analytic solution describing the shift $Q \to \tilde{Q}_{HS}$ from the flat background to the one involving the nonperturbative higher spin configuration in AdS. To make a parallel to the Schnabl's solution for nonperturbative tachyonic background note that, from the on-shell string theory point of view, this solution describes the minimum of the tachyon potential stemming the tachyon's $\beta$-function, computed to all orders of the string perturbation theory. Given the SFT solution for nonperturbative higher-spin background, the cohomology of $\tilde{Q}_{HS}$ would then describe the physical properties of such a background. At the first glance, the structure of such a solution must be enormously complicated. Nevertheless, let's try to imagine its possible structure. The complete fully interacting higher-spin theory in $AdS_d$, no matter how complicated its Lagrangian might be, is largely determined by two objects: structure constants of the higher-spin algebras in $AdS$ and conformal blocks in the dual $CFT_{d-1}$. Moreover, as we shall argue in the next section, as far as the cubic SFT is concerned, for substantially large class of solutions the structure constants of the higher-spin algebra (more precisely, the enveloping of this algebra) alone constitute a sufficient information to control the solutions we are looking for. Thus, if one is able to find a class of SFT solutions determined by the structure constants of the HS algebra, this already would be a strong signal that it describes the higher spin background we are interested in. Our strategy below will thus be the following. First, we shall discuss, as a warm-up example, a set of simple SFT solutions that involve the primary fields only and describe the *perturbative* background deformations. Remarkably, one particular example of these solutions is given by the discrete states in $c = 1$ model where both the structure constants of the $AdS_3$ higher-spin algebra appear and the vertex operators are described in terms of products of the Bell polynomials of the type (5.36). Then we shall develop the OPE formalism for the Bell polynomials of string fields, evaluating their structure constants. We find that these structure constants can be obtained from simple generating function $G(x, y)$ of two variables, which series expansion is determined by coefficients related to $AdS_3$ structure constants. Next, we will propose an ansatz of the form (5.36) solving (5.20). The solution will be described in terms of defining generating function $F(G)$ satisfying certain defining relations derived in the following sections and structurally can be thought of as an envelopping of the higher-spin algebra.

## 5.3 Structure Constants, Higher Spins and SFT Solutions: A Warm-Up Example

One particularly simple and almost obvious example of a class of string fields solving (5.20) can be constructed as follows.

Let $V_i(z,p)(i=1,...)$ be the set of all physical vertex operators in string theory in the cohomology of the original BRST charge $Q$ (primary fields of ghost number 1 and conformal dimension 0) and $\lambda^i(p)$ are the corresponding space-time fields (where $p$ is the momentum in space-time and we suppress the space-time indices for the brevity). Then the string field

$$\Psi_0 = \sum_i \lambda^i V_i \tag{5.36}$$

is the solution of (5.20) provided that the zero $\beta$-function conditions:

$$\beta_{\lambda^i} = 0 \tag{5.37}$$

are imposed on the space-time fields in the leading order of the perturbation theory. This statement is easy to check. Indeed, the on-shell invariance conditions on $V_i$ imply $\{Q, \lambda^i V_i\} = \hat{L}\lambda^i = 0$ where $\hat{L}$ is some differential operator (e.g. a Laplacian plus the square of mass) acting on $\lambda^i$. Next, since the operators are the dimension zero primaries, they are invariant under the transformations (5) and therefore the star product can be computed simply by using

$$<<\Psi, \Psi \star \Psi>>=<\prod_{n=1}^{3} f_n^3 \circ \Psi(0)>=\sum_{i,j,k} C_{ijk}\lambda^i\lambda^j\lambda^k \tag{5.38}$$

where $C_{ijk}(p_1,p_2)$ are the structure constants in front of the simple pole in the OPE of the vertex operators:

$$V_i(z_1,p_1)V_j(z_2,p_2) \sim (z_1-z_2)^{-1}C_{ijk}(p_1,p_2)V^k(\frac{1}{2}(z_1+z_2), p_1+p_2) \tag{5.39}$$

Substitution into the SFT equations of motion then leads to the constraints on $\lambda^i$ space-time fields:

$$\hat{L}\lambda^i + C^i_{jk}\lambda^j\lambda^k = 0 \tag{5.40}$$

which are nothing but $\beta_{\lambda^i} = 0$ equation (5.38) in the leading order. Note that the SFT solution (5.37) is entirely fixed by the leading order contribution to the $\beta$-function (which are completely determined by the 3-point correlation functions of the vertex operators) and does not depend on the higher-order corrections (related to the higher-point correlators). The higher order corrections to the $\beta$-function only appear upon the deformation (5.34) of the BRST charge related to the solution (5.37) which, in this case, simply reduces to $Q \to \tilde{Q} = Q + \sum \lambda_i V^i$. The 4-point functions of the $V_i$ vertex operators will then determine the solution of the equation (5.20) with $Q$ replaced by $\tilde{Q}$. This, in turn, will lead to the further shift of

$\tilde{Q}$ in the next order etc., so the whole procedure can be performed order by order. The physical meaning of these deformations is also quite clear: they define, order by order, the *perturbative* changes of the background caused by the RG flows from the original conformal point (corresponding to flat background) to the new fixed point (corresponding to a certain solution of the low-energy effective equations of motion). Physically, far more interesting is of course the case when the operators entering the string field (5.37) are no longer the primaries of any fixed conformal dimensions and are off-shell, but still solve (5.20) with the constraints of the type (5.41). Then, $\lambda$ describes the background which is beyond the reach of the conventional string perturbation theory while the $C$-constants describe the new $2d$ CFT related to this *non – perturbative* background change. This is precisely the type of the higher-spin related SFT solution we will be looking for. The instructive point here (which follows from the above discussion) is that, if we start with the SFT equation of motion with the unperturbed BRST charge of the bosonic theory:

$$Q = \oint dz\{cT - bc\partial c\} = \oint dz\{-\frac{1}{2}\partial X_m \partial X^m + bc\partial c\} \tag{5.41}$$

the higher-spin solution we are searching for shouldn't depend on higher-point correlators or the conformal blocks, but only on the structure constants of the higher-spin algebra. The final remark we shall make before moving further regards the appearance of the higher spin algebra in the SFT solution of the type (5.37) - (5.41) at the *perturbative* level, as well as the appearance of the Bell polynomials as the operators realizing this algebra. Consider the noncritical open one-dimensional bosonic string theory (also known as the $c = 1$ model). It is well-known that this string theory does not contain a photon in the massless spectrum, however, due to the $SU(2)$ symmetry at the self-dual point, it does contain the $SU(2)$ multiplet of the discrete states which are physical at integer or half-integer momentum values only and become massless upon the Liouville dressing. To obtain the vertex operators for these states, consider the $SU(2)$ algebra generated by

$$T_{\pm} = \oint dz e^{\pm iX\sqrt{2}}$$

$$T_0 = \frac{i}{\sqrt{2}}\partial X \tag{5.42}$$

where $X$ is a single target space coordinate and the dressed BRST-invariant highest weight vector

$$V_l = \int dz e^{(ilX+(l-1)\varphi)\sqrt{2}} \tag{5.43}$$

where $\varphi$ is the Liouville field and $l$ is integer or half-integer. The SU(2) multiplet of the operators is then obtained by repeatedly acting on $V_l$ with the lowering operator $T_-$ of SU(2):

$$U_{l|m} = T_-^{l-m} V_l$$
$$-l \leq m \leq l \tag{5.44}$$

The dressed $U_{l|m}$ operators are the physical operators (massless states) of the $c = 1$ model and are the worldsheet integrals of primary fields of dimension one (equivalently, the primaries of dimension 0 at the unintegrated $b - c$ ghost number 1 picture).

Manifest expressions for the $U_{l|m}$ vertex operators are complicated, however, their structure constants have been deduced by [67, 68] by using symmetry arguments. One has

$$U_{l_1|m_1}(z)U_{l_2|m_2}(w)$$
$$\sim (z-w)^{-1}C(l_1, l_2, l_3|m_1, m_2, m_3)f(l_1, l_2)U_{l_3,m_3} \tag{5.45}$$

where the $SU(2)$ Clebsch-Gordan coefficients are fixed by the symmetry while the function of Casimir eigenvalues $f(l_1, l_2)$ is nontrivial and was deduced to be given by [67, 68]

$$f(l_1, l_2) = \frac{\sqrt{l_1 + l_2}(2l_1 + 2l_2 - 2)!}{\sqrt{2l_1 l_2}(2l_1 - 1)!(2l_2 - 1)!} \tag{5.46}$$

Remarkably, these structure constants coincide (up to a simple field redefinition) exactly with those of $w_\infty$ wedge, defining the asymptotic symmetries of the higher spin algebra in $AdS_3$ in a certain basis, computed in a rather different context by M. Henneaux and S.-J. Rey [149]. Thus the primaries (5.45) are connected to a vertex operator realization of $AdS_3$ higher-spin algebra. The related OSFT solution is then constructed similarly to the previous one. It is given simply by

$$\Psi = \sum_{l,m} \lambda^{l|m} U_{l|m} \tag{5.47}$$

with the constants $\lambda^{l|m}$ satisfying the $\beta$-function condition

$$S^{l_3|m_3}_{l_1 m_1|l_2 m_2} \lambda^{l_1 m_1} \lambda^{l_2 m_2} = 0 \tag{5.48}$$

where $S^{l_3|m_3}_{l_1 m_1|l_2 m_2} = C(l_1, l_2, l_3|m_1, m_2, m_3)f(l_1, l_2)$ are the $AdS_3$ higher spin algebra's structure constants. As previously, this solution describes the perturbative background's change

$$Q \to \tilde{Q} = Q + \sum_{l,m} \lambda^{l|m} U_{l|m} \tag{5.49}$$

The higher order contributions to the $\beta$-function then will appear in the SFT solutions with $Q$ replaced with $\tilde{Q}$ etc. Our particular goal is, roughly speaking, to find the off-shell analogues of the string field (5.36) solving the SFT equation of motion, with $\lambda$-constants satisfying the constraints related to the structures of the higher-spin algebras.

For that, it is first instructive to investigate the manifest form of the operators (note that in [67,68] the structure constants were computed from the symmetry arguments, without pointing out the explicit form of the operators). Taking the highest weight vector $V_l$ and applying $T_-$ using the OPE

$$\begin{aligned} &e^{-iX\sqrt{2}}(z)e^{(ilX+(l-1)\varphi)\sqrt{2}}(w) \\ &= \sum_{k=0}^{\infty}(z-w)^{k-2l}B^{(k)}_{-i\sqrt{2}X} : e^{(i(l-1)X+(l-1)\varphi)\sqrt{2}} : (w) \end{aligned} \tag{5.50}$$

we obtain

$$U_{l|l-1} = T_- V_l = \oint dw : B^{(2l-1)}_{-i\sqrt{2}X} e^{(i(l-1)X+(l-1)\varphi)\sqrt{2}} : (w) \tag{5.51}$$

Here, as previously, $B^{(n)}_{f(z)} \equiv B^{(n)}(\partial_z f, ..., \partial_z^n f)$ are the rank $n$ normalized Bell polynomials in the derivatives of $f$, defined according to

$$\begin{aligned} &B^{(n)}(\partial_z f, ..., \partial_z^n f) = B^{(n)}(x_1, ...x_n)|_{x_k \equiv \partial^k f; 1\leq k\leq n} \\ &= \sum_{k=1}^{n} B_{n|k}(x_1, ...x_{n-k+1}) \end{aligned} \tag{5.52}$$

where $B_{n|k}(x_1, ...x_{n-k+1})$ are the normalized partial Bell polynomials defined according to

$$\begin{aligned} &B_{n|k}(x_1, ...x_{n-k+1}) \\ &= \sum_{p_1,...p_{n-k+1}} \frac{1}{p_1!...p_{n-k+1}!} x_1^{p_1}\left(\frac{x_2}{2!}\right)^{p_2}...\left(\frac{x_{n-k+1}}{(n-k+1)!}\right)^{p_{n-k+1}} \end{aligned} \tag{5.53}$$

with the sum taking over all the combinations of non-negative $p_j$ satisfying

$$\begin{aligned} &\sum_{j=1}^{n-k+1} p_j = k \\ &\sum_{j=1}^{n-k+1} jp_j = n \end{aligned} \tag{5.54}$$

(note that the standard Bell polynomials $P^{(n)}$ are related to the normalized ones as $P^{(n)} = n!B^{(n)}$; similarly for the partial Bell polynomials). To calculate the next vertex operator, $U_{l|l-2} = T_- U_{l|l-1}$ one also needs to point out the OPE between Bell polynomials of the $X$-derivatives and the exponents of $X$ as well. First of all, it is straightforward to deduce the identity

$$B_{\alpha X}^{(n)}(z)e^{\beta X}(w)$$
$$=\sum_{k=0}^{n}(z-w)^{-k}\frac{\Gamma(-\alpha\beta+1)}{k!\Gamma(-\alpha\beta+1-k)}:B_{\alpha X}^{(n-k)}(z)e^{\beta X}(w): \qquad (5.55)$$

where $\alpha$ and $\beta$ are some numbers and $\Gamma$ is the Euler's gamma-function. Note that this is the double point OPE (sufficient for our purposes), i.e. accounting only for the contractions between $B_{\alpha X}^{(n)}$ and $e^{\beta X}(w)$, but not for the expansions of any of them around some fixed point (such as $z$, $w$ or a midpoint).

Next, it is then straightforward to obtain:

$$U_{l|l-2}=2!\oint dw:e^{(i(l-2)X+(l-1)\varphi)\sqrt{2}}$$
$$(B_{-i\sqrt{2}X}^{(2l-1)}B_{-i\sqrt{2}X}^{(2l-3)}-(B_{-i\sqrt{2}X}^{(2l-2)})^2):(w) \qquad (5.56)$$

This operator is given by the exponent multiplied by the quadratic combination of the Bell polynomials with ranks $B^{(2l-k_j)};j=1,2$ with $k_1+k_2$ being the length 2 partition of $2^2=4$ with $1\leq k_{1,2}\leq 2\times 2-1=3$. It is straightforward to continue this sequence of transformations by $T_-$ to identify the manifest expressions for all the vertex operators. For arbitrary $U_{l|l-m}$ $(1\leq m\leq l)$ we obtain

$$U_{l|l-m}=m!\oint dwe^{(i(l-m)X+(l-1)\varphi)\sqrt{2}}$$
$$\sum_{m^2|k_1...k_m}(-1)^{\pi(k_1,...,k_m)}B_{-i\sqrt{2}X}^{(2l-k_1)}B_{-i\sqrt{2}X}^{(2l-k_2)}...B_{-i\sqrt{2}X}^{(2l-k_m)} \qquad (5.57)$$

with the sum taken over all the ordered length $m$ partitions of $m^2=k_1+...+k_m$ with $1\leq k_1\leq....\leq k_m\leq 2m-1$ and with the parity $\pi(k_1,...,k_m)$ of each partition defined as follows. Consider a particular partition of $m^2$: $k_1\leq k_2....\leq k_m$. By permutation we shall call any exchange between two neighbouring elements of the partition with one unit that does not break the order of the partition, e.g.

$$\{k_1\leq...k_{i-1}\leq k_i\leq k_{i+1}\leq k_{i+2}\leq...\leq k_m\}$$
$$\rightarrow\{k_1\leq...k_{i-1}\leq(k_i\pm 1)\leq(k_{i+1}\mp 1)\leq k_{i+2}\leq...\leq k_m\} \qquad (5.58)$$

Then $\pi(k_1,...,k_m)$ for any length $m$ partition of $m^2$ is the minimum number of permutations needed to obtain the partition $m^2=k_1+....+k_m$ from the reference partition $m^2=1+3+5+...+(2m-1)$. Note that, possibly up to an overall sign change of $U_{l|m}$, any partition can be chosen as a reference partition. One particular lesson that we learn from (5.58) is that combinations of the objects of the type $\sum_{\{n_1...n_k\};N=n_1+...+n_k}\alpha_{n_1...n_k}\prod_{j=1}^k B^{(n_j)}$ form a basis for the operator realization of the higher-spin algebra. As we will see below, this is not incidental, as the products of the Bell polynomials naturally realize $w_\infty$ and envelopings of $SU(2)$. In general, they are not primary fields, except for some very special choices of the $\alpha_{n_1...n_k}$ coefficients in the summation over the partitions in (5.58). (Strictly speaking, the products of Bell polynomials in (5.57), (5.58) are the primaries only for $m=l$; otherwise they must be dressed with the exponents.) Two important numbers characterizing these objects are $N$ and $k$ (total conformal dimension and the partition length).

The ansatz for the solution in $D$-dimensional string field theory that we propose is the following. Define the generating function for the Bell polynomials:

$$H(B)=\sum_{n=1}^{\infty}h_nB^{(n)}_{\vec{\alpha}\vec{X}}\equiv\sum_{n=1}^{\infty}\frac{h_nP^{(n)}_{\vec{\alpha}\vec{X}}}{n!}\tag{5.59}$$

where $P^{(n)}$ are the standard (non-normalized) Bell polynomials in the derivatives of $\vec{\alpha}\vec{X}$, $h_n$ are some coefficients, defining the associate characteristic function

$$H(x)=\sum_n\frac{h_nx^n}{n!}\tag{5.60}$$

This function is convenient to use in order to perform various operations with $H(B)$; e.g. the derivative function $H'(B)$ can be obtained by differentiating $H(x)$ over $x$ and then replacing $\frac{x^n}{n!}\to B^{(n)}_{\vec{\alpha}\vec{X}}$ in the expansion series obtained by differentiation. Next, define another characteristic function

$$G(x)=\sum_{n=0}^{\infty}\frac{g_nx^n}{n!}\tag{5.61}$$

Then the composite function $G(H(B))$ generates the products of Bell polynomial operators according to the Faa de Bruno formula (which is easy to check by simple straightforward computation):

$$\begin{aligned}G(H(B))&=\sum_{n=0}^{\infty}\frac{g^n}{n!}\sum_{N=n}^{\infty}N!\\&\times\sum_{N|k_1...k_n}h_{k_1}....h_{k_n}B^{(k_1)}_{\vec{\alpha}\vec{X}}...B^{(k_n)}_{\vec{\alpha}\vec{X}}\sigma^{-1}(k_1,...k_n)\\&=\sum_{n=0}^{\infty}\frac{g^n}{n!}\sum_{N=n}^{\infty}B_{N|n}(h_1B^{(1)}_{\vec{\alpha}\vec{X}},...,h_{N-n+1}B^{(N-n+1)_{\vec{\alpha}\vec{X}}})\end{aligned}\tag{5.62}$$

where $\sum_{N|k_1...k_n}$ stands for the summation over ordered length $n$ partitions of $N$ $(0 < k_1 \leq k_2... \leq k_n)$ and the sigma-factor

$$\sigma(k_1, ..., k_n) = q_{k_1}!...q_{k_n}! \tag{5.63}$$

is the product of the multiplicities of the elements $k_j$ of the partition. (Note that each $q_{k_j}$ elements $k_j$ entering the partition give rise to the single factor of $q_{k_j}!$ in the $\sigma^{-1}$ denominator in (5.63).) We will be looking for the ansatz SFT solution in the form (5.63), that is,

$$\Psi = G(H(B)) \tag{5.64}$$

and our goal is to determine the coefficients $h_n$ and $g_n$ (more precisely, the defining constraints on these coefficients imposed by the SFT equations of motion). $\vec{\alpha}$ is some parameter which a priori is not fixed; however, we shall see below that the star product for $\Psi$ is drastically simplified if the tachyon-like constraint: $\alpha^2 = -2$ is imposed on $\vec{\alpha}$ and it is precisely this simplification that ultimately makes it possible to formulate the SFT solutions in terms of the functional relations for $G(H)$. As it is clear from the above, the SFT ansatz that we propose is given by the series in the partial Bell polynomials of the Bell polynomials in the target space fields. An essential property of these objects is that their operator algebra realizes the enveloping of $SU(2)$ with the enveloping parameter related to $\alpha^2$. In particular, a simple pole in the OPE of these objects leads to classical $w_\infty$ algebra of area-preserving diffeomorphisms, while the complete OPE generates the full enveloping (the explicit OPE structure will be given below). This is where the connection with the higher spin algebra enters the game. To prepare for the analysis of the equations of motion using the ansatz (5.63), we shall first analyze the conformal transformation properties, operator products and the correlators of the vertex operators involving the Bell polynomials and their products.

## 5.4 CFT Properties of SFT Ansatz

The first important building block in our construction of the SFT solution is the analysis of the conformal field theory properties of operators constructed out of products of the Bell polynomials in the target space fields. The first step is to determine the transformation laws for the operators in the sum (5.63). We start from the infinitezimal transformation of a single Bell polynomial. First, we need to evaluate the operator product of the stress tensor with $B^{(n)}_{\vec{\alpha}\vec{X}}$. This can be done by using the identity

$$\partial_z^n e^{\vec{\alpha}\vec{X}}(z) = n! B^{(n)}_{\vec{\alpha}\vec{X}} n e^{\vec{\alpha}\vec{X}} \tag{5.65}$$

Then one can deduce the infinitezimal conformal transformation of $B^{(n)}_{\vec{\alpha}\vec{X}}$ with the generator

$$\oint dz \epsilon(z) T(z) = -\frac{1}{2}\oint dz \epsilon(z) \partial X_m \partial X^m(z) \tag{5.66}$$

by using the identity

$$\delta_\epsilon(\partial_z^n e^{\vec{\alpha}\vec{X}}) = n!(\delta_\epsilon B^{(n)}_{\vec{\alpha}\vec{X}}) e^{\vec{\alpha}\vec{X}} + n! B^{(n)}_{\vec{\alpha}\vec{X}}(\delta_\epsilon e^{\vec{\alpha}\vec{X}}) + \delta_\epsilon(overlap) \tag{5.67}$$

with $\delta_\epsilon(overlap)$ accounting for the contribution in which one of the $\partial X$'s of $T$ is contracted with $B^{(n)}$ and another with the exponent. Using the manifest expression for $B^{(n)}_\psi$:

$$B^{(n)}_\psi = \sum_l \sum_{n|p_1 \dots p_l} \frac{(\partial^{p_1}\psi)^{m_1} \dots (\partial^{p_l}\psi)^{m_l}}{p_1! \dots p_k! m_1! \dots m_k!} \tag{5.68}$$

where $\psi = \vec{\alpha}\vec{X}$ and the sum is taken over the ordered partitions

$$\begin{aligned} n &= \sum_{j=1}^{l} m_j p_j \\ k &= \sum_{j=1}^{l} m_j \\ 1 &\leq k \leq n; 1 \leq l \leq k \\ p_1 &< p_2 < \dots < p_l \end{aligned} \tag{5.69}$$

it is straightdorward to establish the OPE:

$$\partial X_m(z) B^{(n)}_\psi(w) = -\alpha_m \sum_{k=1}^{n} (z-w)^{-k-1} B^{(n-k)}_\psi(w) + regular \tag{5.70}$$

Using this OPE it is straightforward to compute the overlap transformation and to deduce the OPE between the stress-energy tensor and $B^{(n)}_\psi$ with the result given by

$$\begin{aligned} T(z) B^{(n)}_\psi(w) &= (z-w)^{-1} \partial B^{(n)}_\psi(w) + n(z-w)^{-2} B^{(n)}_\psi(w) + \\ &+ \sum_{k=2}^{n+1} (z-w)^{-k-1} (n+1+\alpha^2 - \frac{1}{2}(\alpha^2+2)k) B^{(n-k+1)}_\psi(w) \end{aligned} \tag{5.71}$$

Note that the coefficients in front of $B_\psi^{(n-k+1)}(w)$ do not depend on $k$ when $\vec{\alpha}$ satisfies the tachyon-like condition $\alpha^2 = -2$. This drastically simplifies the problem to determine the behaviour of $B^{(n)}$ under the finite conformal transformations in the SFT equations, which infinitezimal form is defined by the OPE (5.72). It is now straightforward to deduce the OPE of $T(z)$ with the product of any number $q$ of the Bell polynomials of the target space fields, which will be the main building block for the SFT solutions that we are looking for. It is convenient to introduce the notation:

$$R_N^{n_1 \dots n_q} = \prod_{j=1}^{q} B_\psi^{(n_j)}(w)$$
$$N = \sum_j n_j \tag{5.72}$$

We have:

$$\begin{aligned}
&T(z)R_N^{n_1\dots n_q}(w)\\
&= \sum_{j=1}^{q}\sum_{k_j=2}^{n_j+1}(z-w)^{-k_j-1}(n_j+1+\alpha^2-\frac{1}{2}(\alpha^2+2)k_j)\\
&\times R_{N-n_j}^{n_1\dots n_q|_j}B_\psi^{(n_j-k_j+1)}\\
&-\alpha^2\sum_{l,m=1;l<m}^{q}(z-w)^{-k_l-k_m}(R_{N-n_l-n_m}^{n_1\dots n_q|_{l,m}})B_\psi^{(n_l-k_l+1)}B_\psi^{(n_m-k_m+1)}\\
&+N(z-w)^{-2}R_N^{n_1\dots n_q}(w)+(z-w)^{-1}\partial R_N^{n_1\dots n_q}(w)
\end{aligned} \tag{5.73}$$

where $n_1 \dots n_q|_j$ stands for the set of $q-1$ indices with $n_j$ excluded, similarly for $n_1 \dots n_q|_{l,m}$. Now it is straightforward to obtain the infinitezimal thc transformation law for $cR_N^{n_1\dots n_q}$:

We have:

$$\begin{aligned}
&\delta_\epsilon(cR_N^{n_1\dots n_q})(w)\\
&= \sum_{j=1}^{q}\sum_{k_j=2}^{n_j+1}\frac{\partial^{k_j}\epsilon}{k_j!}(n_j+1+\alpha^2-\frac{1}{2}(\alpha^2+2)k_j)cR_{N-n_j}^{n_1\dots n_q|_j}B_\psi^{(n_j-k_j+1)}\\
&-\alpha^2\sum_{l,m=1;l<m}^{q}\frac{\partial^{k_l+k_m-1}\epsilon}{(k_l+k_m-1)!}cR_{N-n_l-n_m}^{n_1\dots n_q|_{l,m}}B_\psi^{(n_l-k_l+1)}B_\psi^{(n_m-k_m+1)}\\
&+(N-1)\partial\epsilon R_N^{n_1\dots n_q}(w)+\epsilon\partial(cR_N^{n_1\dots n_q})(w)
\end{aligned} \tag{5.74}$$

Now we have to establish transformation law for $cR_N^{n_1\ldots n_q}$ under $z \to f(z)$, necessary to compute the correlators in the string field theory equations of motion. This can be deduced from two conditions: first, the finite transformation should reproduce the infinitezimal one for $f(z) = z + \epsilon$. Second, the form of the global transformation must be preserved under the composition. As it is well-known, in case of simplest non-primary field, such as the stress-energy tensor, this leads to the appearance of the Schwarzian derivative of $f(z)$ which is in fact the degree 2 Bell polynomial in $log(f'(z))$:

$$S(f(z)) = 2B_2(-\frac{1}{2}log(f')) \tag{5.75}$$

This is not a coincidence since for a large class of non-primaries in $CFT$ the higher degree Bell polynomials correspond to the higher derivative extensions of the Schwarzian derivative in conformal transformations. Note that the Bell polynomials of the logarithms of functions defining global conformal transformation satisfy the following composition identity:

$$B^{(n)}(log(\frac{d}{dx}f(g(x)))) = \sum_k B^{(n-k)}(log(f'(g)))B^{(k)}(log(g'(x))) \tag{5.76}$$

making them natural objects present in global conformal transformations. The finite conformal transformations of the Bell polynomial operators, consistent with their infinitezimal transformations, are deduced to be given by

$$\begin{aligned}
&cR_N^{n_1\ldots n_q}(z) \to_{z\to f(z)} (\frac{df}{dz})^{N-1}cR_N^{n_1\ldots n_q}(f(z))\\
&-\sum_{j=1}\sum_{k_j=2}^{n_j+1}\frac{1}{k_j}(\frac{df}{dz})^{N-k_j}\\
&\times B^{(k_j-1)}(-(n_j+1+\alpha^2-\frac{1}{2}(\alpha^2+2)k_j)log(\frac{df}{dz}))\\
&\times cR_{N-n_j}^{n_1\ldots n_q|_j}B_\psi^{(n_j-k_j+1)}\\
&+\sum_{l,m=1;l<m}^{q}\frac{1}{(k_l+k_m-1)}(\frac{df}{dz})^{N-k_l-k_m+1}\\
&\times B_{k_l+k_m-2}(-\alpha^2 log(\frac{df}{dz}))\\
&\times cR_{N-n_l-n_m}^{n_1\ldots n_q|_{l,m}}B_\psi^{(n_l-k_l+1)}B_\psi^{(n_m-k_m+1)}
\end{aligned} \tag{5.77}$$

For our purposes, we shall need to compute the values of the Bell polynomials in the transformation law (5.75) for the functions $I(z) = -\frac{1}{z}$ (in the kinetic term of the SFT action) and $g \circ f_k^3(z)$ at $z = 0$ where, as before, $f_k^3$ defined in (5.18) map the string worldsheets of the cubic theory to the wedges of the disc and

$$g(z) = i\frac{1-z}{1+z} \tag{5.78}$$

further maps this disc to the half-plane, so that

$$\begin{aligned} g \circ f_1^3(0) &= 0 \\ g \circ f_2^3(0) &= \sqrt{3} \\ g \circ f_3^3(0) &= -\sqrt{3} \end{aligned} \tag{5.79}$$

For that, we shall use the fact that if $\{a_n\}$ are the coefficients in the series expansion of any function $f(x) = \sum_{n=1} \frac{a_n x^n}{n!}$ (assume $f(0) = 0$), then $e^{f(x)} = 1 + \sum_{n=1}^{\infty} B^{(n)}(a_1, ..., a_n)x^n$. From now on, to abbreviate things, we shall restrict ourselves to the case $\alpha^2 = -2$, relevant to our SFT solution. We start from $B^{(k)}(\kappa log(I'(z)))$ where, in particular, $\kappa = 1 - n_j$ for the terms in the first sum in (5.77) and $\kappa = -\alpha^2 = 2$ is in the second. Then

$$B^{(n)}(\kappa log(I'(z))) = z^{-n}B^{(n)}(2\kappa, -2\kappa, ...(-1)^n 2(n-1)!\kappa) \tag{5.80}$$

The Bell polynomial on the right-hand side is then identified with the $n$'th expansion coefficient of the exponent of $-2\kappa log z$, i.e. of $z^{-2\kappa}$ Therefore

$$B^{(n)}(\kappa log(I'(z))) = \frac{\Gamma(1-2\kappa)z^{-n}}{n!\Gamma(1-2\kappa-n)} \tag{5.81}$$

Next, we need the values of the Bell polynomials $B^{(n)}(log\frac{df(z)}{dz})$ with $f(z) = g \circ f_k^3(z)$ at $z = 0$. Straightforward calculation gives the result

$$B^{(n)}(\kappa log\frac{df(z)}{dz})|_{z=0} = B^{(n)}(\beta_1 ... \beta_k ... \beta_n) \tag{5.82}$$

with

$$\beta_k = \kappa(\frac{5}{3}(-i)^n + \frac{1}{3}(i)^n - (\frac{1}{2})^{n-1})(n-1)! \tag{5.83}$$

for $g \circ f_1^3(z)$,

$$\beta_k = \kappa(\frac{5}{3}(-i)^n + \frac{1}{3}(i)^n + 2e^{\frac{2i\pi n}{3}})(n-1)! \tag{5.84}$$

for $g \circ f_2^3(z)$ and

$$\beta_k = \kappa(\frac{5}{3}(-i)^n + \frac{1}{3}(i)^n + 2e^{\frac{-2i\pi n}{3}})(n-1)! \tag{5.85}$$

for $g \circ f_3^3(z)$. Accordingly, these Bell polynomials are identified with the expansion series of

$$h_1(z) = (1+iz)^{-\frac{5}{3}\kappa}(1-iz)^{-\frac{1}{3}\kappa}(1+\frac{z}{2})^{-2\kappa}$$
$$h_2(z) = (1+iz)^{-\frac{5}{3}\kappa}(1-iz)^{-\frac{1}{3}\kappa}(1-e^{\frac{2i\pi}{3}}z)^{-2\kappa}$$
$$h_3(z) = (1+iz)^{-\frac{5}{3}\kappa}(1-iz)^{-\frac{1}{3}\kappa}(1-e^{\frac{-2i\pi}{3}}z)^{-2\kappa} \quad (5.86)$$

for $g \circ f_1^3$, $g \circ f_2^3$ and $g \circ f_3^3$ respectively. Accordingly, the values of the Bell polynomials are given by

$$B^{(n)}(\kappa log(\frac{d}{dz}g \circ f_1^3(z)))|_{z=0}$$
$$= \sum_{k,l,m|k+l+m=n} \frac{e^{\frac{i\pi}{2}(k-l)}2^{-m}}{k!l!m!}$$
$$\times \frac{\Gamma(1-\frac{5}{3}\kappa)\Gamma(1-\frac{1}{3}\kappa)\Gamma(1-2\kappa)}{\Gamma(1-\frac{5}{3}\kappa-k)\Gamma(1-\frac{1}{3}\kappa-l)\Gamma(1-2\kappa-m)}$$
$$B^{(n)}(\kappa log(\frac{d}{dz}g \circ f_2^3(z)))|_{z=0}$$
$$= \sum_{k,l,m|k+l+m=n} \frac{e^{i\pi(\frac{1}{2}(k-l)+\frac{2m}{3})}}{k!l!m!}$$
$$\times \frac{\Gamma(1-\frac{5}{3}\kappa)\Gamma(1-\frac{1}{3}\kappa)\Gamma(1-2\kappa)}{\Gamma(1-\frac{5}{3}\kappa-k)\Gamma(1-\frac{1}{3}\kappa-l)\Gamma(1-2\kappa-m)}$$
$$B^{(n)}(\kappa log(\frac{d}{dz}g \circ f_3^3(z)))|_{z=0}$$
$$= \sum_{k,l,m|k+l+m=n} \frac{e^{i\pi(\frac{1}{2}(k-l)-\frac{2m}{3})}}{k!l!m!}$$
$$\times \frac{\Gamma(1-\frac{5}{3}\kappa)\Gamma(1-\frac{1}{3}\kappa)\Gamma(1-2\kappa)}{\Gamma(1-\frac{5}{3}\kappa-k)\Gamma(1-\frac{1}{3}\kappa-l)\Gamma(1-2\kappa-m)} \quad (5.87)$$

with the sums taken over the unordered partitions of $n = k+l+m$. These relations altogether fully determine the transformation properties of our string field ansatz, including the star product.

The final step to make before actually computing the SFT correlators is to point out the operator product rules involving the Bell polynomial operators and their blocks. We will do this in the next section, in particular deriving an analogue of the generalized Wick's theorem for the Bell polynomial operators and pointing the relevance of their correlators to the structure constants of the higher-spin algebra.

## 5.5 Bell Polynomial Operators: Operator Products and Correlators

The most crucial building block in our computations involves the OPE rules for the operators of the SFT ansatz of the form (5.63) which we will establish in this section. Ultimately, it turns out that it is precisely the structure of these OPE rules which makes it possible to work out the SFT solution and, moreover, to relate it to the higher spin algebra.

We start from the simplest OPE between $B^{(N)}_{\vec{\alpha}\vec{X}}(z)$ and$B^{(m)}_{\vec{\beta}\vec{X}}(w)$. This doesn't turn out to be an easy OPE to compute. The manifest expressions for the Bell polynomials do not appear to be very helpful. Nevertheless, there are some observations to simplify the computation. First of all, the OPE has to preserve the conformal transformation structure of the Bell polynomial operators. This suggests that the OPE must have the structure

$$\begin{aligned}&B^{(N)}_{\vec{\alpha}\vec{X}}(z)B^{(M)}_{\vec{\beta}\vec{X}}(w)\\&=\sum_{n=0}^{N}\sum_{m=0}^{M}(z-w)^{-n-m}\lambda^{N|M}|_{m|n}:B^{(N-n)}_{\vec{\alpha}\vec{X}}(z)B^{(M-m)}_{\vec{\beta}\vec{X}}(w): \end{aligned}\quad (5.88)$$

(again, for the brevity we consider the double point OPE here, just as was explained above). In other words, the Bell polynomial structure of the operators is preserved by (5.89). The next helpful hint comes from the identity (5.66) relating the Bell operators to the derivatives of the exponents and from analyzing the correlator

$$\begin{aligned}&< B^{(N)}_{\vec{\alpha}\vec{X}}e^{\vec{\alpha}\vec{X}}(z)B^{(M)}_{\vec{\beta}\vec{X}}e^{\vec{\beta}\vec{X}}(w)\\&=\frac{1}{N!M!}\partial_z^N\partial_w^M< e^{\vec{\alpha}\vec{X}}(z)e^{\vec{\beta}\vec{X}}(w)>\\&=(z-w)^{-\vec{\alpha}\vec{\beta}-N-M}\frac{\Gamma(1-\vec{\alpha}\vec{\beta})}{N!M!\Gamma(1-\vec{\alpha}\vec{\beta}-M-N)}\end{aligned}\quad (5.89)$$

This correlator can be computed in two equivalent ways: one either starts with applying the operator algebra of the Bell polynomials with the exponents and then contracting the remaining derivatives of $X$ between themselves in each of the OPE terms - or, alternatively, starting with the operator product between the Bell polynomials, containing the unknown $\lambda$-constants and then contracting the remaining derivatives of X in each of the operators with the opposite exponent. Comparison of these two expressions identifies the remarkably simple OPE structure:

$$B^{(N)}_{\vec{\alpha}\vec{X}}(z)B^{(N)}_{\vec{\beta}\vec{X}}(w)$$
$$=< B^{(n)}_{\vec{\alpha}\vec{X}}(z)B^{(m)}_{\vec{\beta}\vec{X}}(w) >: B^{(N-n)}_{\vec{\alpha}\vec{X}}(z)B^{(M-m)}_{\vec{\beta}\vec{X}}(w) : \quad (5.90)$$

i.e. the OPE coefficients are simply given by the two-point correlators of the lower rank polynomials:

$$\lambda^{N|M}|_{m|n} \equiv \lambda_{m|n} = (z-w)^{n+m} < B^{(n)}_{\vec{\alpha}\vec{X}}(z)B^{(m)}_{\vec{\beta}\vec{X}}(w) > \quad (5.91)$$

The last step is to compute the two-point correlators and somehow this again doesn't turn out to be an elementary exercise. Straightforward calculation using the manifest expression for the Bell polynomials and the Wick's theorem leads to complicated sum over partitions which doesn't seem to be realistic to evaluate and doesn't look illuminating or useful for our purposes. Instead, we shall start from the identity

$$B^{(n)}_{\vec{\alpha}\vec{X}} = \frac{1}{n}(\partial B^{(n-1)}_{\vec{\alpha}\vec{X}} + \vec{\alpha}\partial\vec{X}B^{(n-1)}_{\vec{\alpha}\vec{X}}) \quad (5.92)$$

Inserting this identity in the correlator (5.92) and using the OPE (5.89) we obtain the recursion relation

$$\lambda_{n|m} = -\frac{n+m-1}{n}\lambda_{n-1|m} - \frac{\vec{\alpha}\vec{\beta}}{n}\sum_{l=1}^{m-1}\lambda_{n-1|l} \quad (5.93)$$

This recursion relation can be simplified by repeating the above procedure and inserting the identity (5.93) into the correlator $< B^{(n)}B^{(m-1)} >$, obtaining the similar recursion relation for $\lambda_{n|m-1}$ and subtracting it from (5.94). Then the recursion becomes

$$n(\lambda_{n|m} - \lambda_{n|m-1}) = -(n+m-1)\lambda_{n-1|m}$$
$$+(n+m-2-\vec{\alpha}\vec{\beta})\lambda_{n-1|m-1} \quad (5.94)$$

with the obvious physical constraints

$$\lambda_{0|k} = \lambda_{k|0} = \delta_{0k} \quad (5.95)$$

To solve this recursion, define the generating function

$$F_\lambda(x,y) = \sum_{m,n}\lambda_{n|m}x^n y^m, \quad (5.96)$$

multiply the recursion (5.94) by $x^n y^m$ and sum over $m$ and $n$. This leads to the first order partial differential equation for $F_\lambda(x,y)$:

$$(1-y)(1+x)\partial_x F_\lambda + y(1-y)\partial_y F_\lambda + \vec{\alpha}\vec{\beta}yF = 0 \quad (5.97)$$

with the boundary conditions

$$F_\lambda(x,0) = F_\lambda(0,y) = 1 \tag{5.98}$$

This equation isn't hard to solve. Defining

$$\xi = log(1+x), \eta = log(y)$$
$$G(x,y) = logF(x,y) \tag{5.99}$$

the equation simplifies according to

$$\partial_\xi G(\xi,\eta) + \partial_\eta G(\xi,\eta) - \frac{\vec{\alpha}\vec{\beta}y}{1-e^{-\eta}} = 0 \tag{5.100}$$

and is equivalent to the characteristic ODE system

$$\frac{d\xi}{ds} = \frac{d\eta}{ds} = 1$$
$$\frac{dG}{ds} = \frac{\vec{\alpha}\vec{\beta}y}{1-e^{-\eta}} \tag{5.101}$$

so the general solution is

$$G(\xi,\eta) = H(\xi-\eta) + \vec{\alpha}\vec{\beta}\int\frac{d\eta}{1-e^{-\eta}} \tag{5.102}$$

Substituting $G(0,\eta) = 0$ then fixes $H$ to be

$$H(\xi-\eta) = \vec{\alpha}\vec{\beta}log(1-e^{\eta-\xi}) \tag{5.103}$$

so the solution is

$$F_\lambda(x,y) = (\frac{(1+x)(1-y)}{(1+x-y)})^{-\vec{\alpha}\vec{\beta}}$$
$$\lambda_{n|m} = \frac{1}{n!m!}\partial_x^n\partial_y^m F_\lambda(x,y)|_{x,y=0} \tag{5.104}$$

This solution, describing the correlator of two Bell polynomial operators is related to the higher-spin algebra in $AdS_3$ and determines the parameter $\mu$ of the enveloping $T(\mu)$ of $SU(2)$ [67–69].

Now, using the operator products (5.89), (5.90), it is straightforward to identify the worldsheet correlators of the products of the Bell operators in terms of the $\lambda_{m|n}$ -numbers, relevant to $w_\infty$ and to the SU(2) enveloping generators, as well as to our string field theory ansatz solution (5.63). The result is given by

$$<:B_{\vec{\alpha}}^{(n_1)}...B_{\vec{\alpha}}^{(n_p)}:(z)$$
$$:B_{\vec{\beta}}^{(m_1)}...B_{\vec{\beta}}^{(m_q)}:(w)>|_{N=n_1+...+n_p;M=m_1+...+m_q}$$
$$=(z-w)^{-N-M}\sum_{partitions[\alpha_{ij},\beta_{ji}]}\prod_{i=1}^{p}\prod_{j=1}^{q}(q!)^p\lambda_{\alpha_{ij}|\beta_{ji}}$$
$$(\prod_{k=1}^{q}\sigma(\alpha_{1k}|\beta_{k1})!)^{-1}...(\prod_{k=1}^{q}\sigma(\alpha_{pk}|\beta_{kp})!)^{-1}(\prod_{l=1}^{p}\sigma(s_l)!)^{-1} \tag{5.105}$$

with the constraints

$$\sum_{j=1}^{q}\alpha_{ij}=n_i$$
$$\sum_{i=1}^{q}\beta_{ji}=m_j$$
$$\sum_{i=1}^{p}\alpha_{ij}=r_j$$
$$\sum_{j=1}^{q}\beta_{ji}=s_i$$
$$\sum_{j=1}^{q}r_j=\sum_{i=1}^{p}n_i=N$$
$$\sum_{j=1}^{q}m_j=\sum_{i=1}^{p}s_i=N \tag{5.106}$$

with the notations as follows. We have introduced the *exchange numbers* $\alpha_{ij}$ indicating how much of the total conformal dimension $n_i$ of the $B^{(n_i)}(z)$-operator in the product of the Bell polynomials on the left at $z$ is contributed to its interaction with the operator $B^{(m_j)}(w)$ in the product of the Bell polynomials at $w$ on the right, according to the OPE structure (5.89). Similarly, the exchange number $\beta_{ji}$ indicates the reduction in the conformal dimension of $B^{(m_j)}(w)$ on the right as a result of its interaction to $B^{(n_i)}(z)$ on the left. Altogether, this corresponds to the order of $(z-w)^{-\alpha_{ij}-\beta_{ji}}$ term in the OPE of these two operators entering the left and the right chains, contributing to the overall correlator. Thus $r_j$-numbers, forming the length $q$ partition of $N$ (as opposed to the length $p$ partition of $N$, formed by $n_i$) indicate the total loss of conformal dimension of the complete operator on the left-hand side at $z$ due to the interaction with the single polynomial $B^{(m_j)}(w)$ on the right. Similarly, the $s_i$-numbers, forming the length $p$ partition of $M$ (as opposed to the length $q$ partition of $M$ formed by $m_i$), indicate the total loss of conformal dimension of the complete operator on the right-hand side at $w$ due to the interaction with the single polynomial $B^{(n_i)}(z)$ on the left. Next, $\sigma(\alpha_{jk}|\beta_{kj})(j=1,...,p)$ indicates the multiplicity of the array of the exchange numbers $\alpha_{jk}|\beta_{kj}$ in $p$ arrays of the length $q$ each: $\{\alpha_{j1}|\beta_{1j},...,\alpha_{jq}|\beta_{qj}\}$ $(j=1,...p)$, similarly to (34). Finally, $\sigma(s_l)$ counts multiplicities of the $s$-numbers defined above. As before, all the partitions are considered ordered. While the sum (5.106) involving the products of the exchange numbers, summed over the partitions, looks tedious, there are some significant simplifications in important cases, when the partitions are summed over.

Let us again start with the simplest possible warm-up example of summing over the partitions - with all the partition elements summed over uniformly, that is, with the sum being a Bell polynomial of Bell polynomials. Namely, consider the elementary example of a toy string field given by

$$\Psi = \sum_{M=0}^{\infty}\sum_{q=1}^{M}\sum_{M|m_1...m_q}(\prod_{i=1}^{q}[\sigma(m_i)]^{-1})B^{(m_1)}....B^{(m_q)} \qquad (5.107)$$

and let's calculate the simplest SFT correlator $<\Psi(1)\Psi(0)>$. To calculate this correlator, the expression (5.107) must be further summed over the partitions according to the definition of the toy $\Psi$. Let us take the the product (5.107), take the correlator $<B^{(n_1)}....B^{(n_p)}(1)B^{(m_1)}....B^{(m_q)}(0)>$ and begin with the summation over partitions in the second operator at 0. Consider the first row in this product:

$$\sim \sum_{partitions(n_1,s_1)}\prod_{j=1}^{q}(q!)\lambda_{\alpha_{ij}|\beta_{ji}}(\prod_{k=1}^{q}\sigma(\alpha_{1k}|\beta_{k1}))^{-1} \qquad (5.108)$$

with the sum taken over the partitions of $n_1$ and $s_1$ into the exchange number sets. This row completely describes the interaction of $B^{(n_1)}$ with the array of the Bell polynomials at $w$ with $M$ being the total conformal dimension of the array. Let us calculate the effect of the partition summation (5.107) for this row. Now, in addition to the summation over the above partitions, sum over all the partitions of $M$ with lengths $1 \leq q \leq M$ and uniform weights for each $q$.

It is then straightforward to check that the result will be given by the series expansion coefficient of the following simple generating function:

$$\sim \sum_{partitions(n_1,s_1)}\prod_{j=1}^{q}(q!)\lambda_{\alpha_{ij}|\beta_{ji}}$$

$$(\prod_{k=1}^{q}\sigma(\alpha_{1k}|\beta_{k1})!)^{-1}$$

$$= \frac{1}{n_1!s_1!}\partial_x^{n_1}\partial_y^{s_1}(\frac{1}{1-F_\lambda(x,y)})|_{x,y=0} \qquad (5.109)$$

The same procedure can be repeated for the remaining $p-1$ rows parametrized by $(n_j, s_j), j = 1, ...p$, leading to the

$$\sim \frac{\partial_x^{n_1}\partial_y^{s_1}(\frac{1}{1-F_\lambda(x,y)})...\partial_x^{n_p}\partial_y^{s_p}(\frac{1}{1-F_\lambda(x,y)})}{n_1!s_1!...n_p!s_p![\sigma(n_1|s_1)]!...[\sigma(n_p|s_p)]!}$$

Finally, let us sum over the partitions for the first string field at 1. For the fixed values of $N$ and $M$ the result is

$$\sim \frac{1}{M!N!}\partial_x^N\partial_y^M(\frac{1}{1-\frac{1}{1-F_\lambda}})|_{x,y=0} = -\partial_x^N\partial_y^M\frac{1}{F_\lambda}|_{x,y=0}, \tag{5.110}$$

so the two-point function of the toy string field (78) is

$$<\Psi(1)\Psi(0)>= e^{-\frac{1}{F_\lambda(x,y)}}|_{x=y=1} \tag{5.111}$$

The objects of the type (78) are of interest to us both because they are relevant to our SFT ansatz and, at the same time, form an operator algebra realization of $w_\infty$ and $SU(2)$ envelopings. Namely, instead of the string field $\Psi$ (78) consider the field

$$\Psi_{N|p} = \sum_{q=1}^{p}\sum_{N|n_1...n_q}(\prod_{i=1}^{q}[\sigma(n_i)]^{-1})B^{(n_1)}....B^{(n_q)} \tag{5.112}$$

(it is easy to see that the toy string field $\Psi$ is given by $\Psi = \sum_{N=0}^{\infty}\Psi_{N|N}$). This field is characterized by the numbers $N$ and $p$, with the first being its total conformal dimension and the second indicating the maximum length of the "words" made out of Bell polynomial "letters", contained in the string field "sentence" $\Psi_{N|p}$. Let us compute the OPE of two "sentences" $\Psi_{N_1|p_1}(z)$ and $\Psi_{N_2|p_2}(w)$ around the midpoint $\frac{1}{2}(z+w)$. Clearly, the conformal transformation properties of the Bell polynomials imply that the Bell polynomial structure must be preserved under such an operator product. It is also clear from the OPE structures for the Bell polynomials that the terms ("sentences")of the order $(z-w)^{-N}(N>0)$ would consist of words of conformal dimension $N_1+N_2-N$ and lengths up to $p_1+p_2$. For $N=1$ this sends a strong hint towards the emergence of $w_\infty$ and of $SU(2)$ envelopings for higher order $N$ values. Indeed, straightforward calculation leads to the following midpoint OPE simple pole:

$$\Psi_{N_1|p_1}(z)\Psi_{N_2|p_2}(w)$$
$$= (z-w)^{-1}(N_2p_1 - N_1p_2)\Psi_{N_1+N_2-1|p_1+p_2}(\frac{z+w}{2}) \tag{5.113}$$

and the general OPE structure:

$$\begin{aligned}
&\Psi_{N_1|p_1}(z)\Psi_{N_2|p_2}(w) \\
&= (z-w)^{-1}(N_2p_1 - N_1p_2)\Psi_{N_1+N_2-1|p_1+p_2}(\frac{z+w}{2}) \\
&+ \sum_{n=2}^{N_1+N_2}(z-w)^{-n}\gamma_n(N_1,p_1|N_2,p_2)\Psi_{N_1+N_2-n|p_1+p_2}(\frac{z+w}{2})
\end{aligned} \tag{5.114}$$

Although we have not computed the $\gamma_n$ coefficients explicitly, such a computation doesn't look a conceptual challenge and the result must be anyway determined by combinations of the $\lambda$-numbers (5.93) stemming from the two-point correlators of the Bell polynomials. So we recognize classical $w_\infty$ at the simple pole and the enveloping $T(\mu)$ of SU(2) at the higher order singularities with the $\mu$-parameter related to the $\lambda$-numbers. Note that this is the midpoint OPE. If, for example, one needs to compute the OPE around the $w$-point, the right-hand side of (5.114) must be shifted from $\frac{1}{2}(z+w)$ to $w$ by the appropriate series expansion in $\frac{1}{2}(z-w)$. This way, the full enveloping algebra will appear, for example, in commutators of the charges $\oint dz\Psi_{N|n}$.

Now consider a more general example of a string field, relevant to our ansatz. Consider again a generating function of the normalized Bell polynomials:

$$H(B) = \sum_{n=0}^{\infty}\frac{h_n}{n!}B^{(n)}_{\vec{\alpha}\vec{X}}$$

and the string field given by

$$\Psi = G \circ H(B) \tag{5.115}$$

where

$$G(x) = \sum_{n=0}^{\infty}\frac{g_n}{n!}x^n$$

with some fixed coefficients $h_n$ and $g_n$. Using the OPE rules and the formalism developed above, the calculation of the two-point function gives:

$$< \Psi(1)\Psi(0) >=$$
$$\sum_{N=0}^{\infty}\sum_{M=0}^{\infty}\sum_{n=0}^{N}\sum_{m=0}^{M} g_n g_m$$
$$\sum_{N|p_1...p_n}\sum_{M|q_1...q_m}\frac{h_{p_1}...h_{p_n}h_{q_1}...h_{q_m}}{p_1!...p_n!q_1!...q_m!\prod_{i,j}[\sigma(p_i)]![\sigma(q_j)]!}$$
$$\times \sum_{partitions[\alpha_{ij},\beta_{ji}]}\prod_{i=1}^{n}\prod_{j=1}^{m}\lambda_{\alpha_{ij}|\beta_{ji}}(\prod_{k=1}^{m}\sigma(\alpha_{1k}|\beta_{k1})!)^{-1}\times...$$
$$... \times (\prod_{k=1}^{m}\sigma(\alpha_{nk}|\beta_{kn})!)^{-1}(\prod_{l=1}^{n}\sigma(s_l)!)^{-1} \tag{5.116}$$

with the constraints

$$\sum_{j=1}^{m}\alpha_{ij} = n_i; \sum_{i=1}^{n}\beta_{ji} = m_j$$
$$\sum_{i=1}^{n}\alpha_{ij} = r_j; \sum_{j=1}^{m}\beta_{ji} = s_i$$
$$\sum_{j=1}^{m} r_j = \sum_{i=1}^{n} n_i = N$$
$$\sum_{j=1}^{m} m_j = \sum_{i=1}^{n} s_i = N \tag{5.117}$$

with the multiplicity $\sigma$-factors, exchange numbers and $r$, $s$-numbers defined as before. By direct comparison, it is straightforward to realize that the lengthy expression on the right-hand side of (5.116), (5.117) is just a series expansion of the relatively simple generating composite function, that is, it can be cast as simply as

$$< \Psi(1)\Psi(0) >= \tilde{G}(H(F_\lambda(x,y)); H(F_\lambda(x,y)))|_{x=y=1} \tag{5.118}$$

where the function of two variables $\tilde{G}(x,y)$ is related to the function $G(x)$ with the single argument according to

$$\tilde{G}(x,y) = \sum_{m,n}\frac{g_m g_n}{m!n!}x^m y^n \tag{5.119}$$

where $g_n$ are the expansion coefficients of $G$. It is instructive to generalize this two-point correlator to the case of two different string fields, that is, for the case of string fields of the type (5.115) with the different $H$-functions, but with the same $G$-function. The calculation, completely similar to the above, gives:

$$\Psi_1 = G \circ H_1(B)$$
$$\Psi_2 = G \circ H_2(B)$$
$$< \Psi_1(1)\Psi_2(0) >= \tilde{G}(H_1(F_\lambda(x,y)); H_2(F_\lambda(x,y)))|_{x=y=1} \qquad (5.120)$$

The next step in the computation the SFT correlators relevant to the equations of motion in SFT is to determine how the global conformal transformations by $I(z)$ and $g_k(z) \equiv g \circ f_k^3(z)$ act on the string fields of the type (5.115). Using the transformations (5.18), it is not difficult to deduce that, under any of these conformal transformations (denoted by $f(z)$ for the brevity) the string field (5.115) transforms as

$$\Psi \equiv G(H(B)) \to \hat{f}\Psi$$
$$= \frac{dG}{dH}\sum_{n=1}^{\infty}\sum_{k=1}^{n}\alpha_f(k,n)[(B-1)\partial_B^k H(B) + \partial_B^{k-1}H(B)]$$
$$+\frac{d^2G}{dH^2}\sum_{m,n=1;m<n}^{\infty}\ \sum_{k,l=1;m<n}^{k+l=n-1}\partial^k H\partial^l H\beta_f(k,l|n) \qquad (5.121)$$

with the differentiation rules for $G$ and $H$ explained above and with the coefficients $\alpha_f$ and $\beta_f$ related to the conformal transformations by $I(z) = -\frac{1}{z}$ and $g \circ f_k^3(z)$ according to:

$$\alpha_{I(z)}(k,n) = \frac{1}{k!}\frac{\Gamma(2n-1)}{\Gamma(2n-k)}$$
$$\alpha_{g\circ f_1^3(z)}(k,n) = \frac{1}{k}(-\frac{2}{3})^{n-k}\sum_{a,b,c|a+b+c=k-1}\frac{e^{\frac{i\pi}{2}(a-b)}2^{-c}}{a!b!c!}$$
$$\times\frac{\Gamma(1-\frac{5}{3}(1-n))\Gamma(1-\frac{1}{3}(1-n))\Gamma(1-2(1-n))}{\Gamma(1-\frac{5}{3}(1-n)-a)\Gamma(1-\frac{1}{3}(1-n)-b)\Gamma(1-2(1-n)-c)}$$
$$\alpha_{g\circ f_2^3(z)}(k,n) = \frac{1}{k}(-\frac{8}{3})^{n-k}\sum_{a,b,c|a+b+c=k-1}\frac{e^{(\frac{i\pi}{2}(a-b)+\frac{2i\pi c}{3})}2^{-c}}{a!b!c!}$$

$$\times\frac{\Gamma(1-\frac{5}{3}(1-n))\Gamma(1-\frac{1}{3}(1-n))\Gamma(1-2(1-n))}{\Gamma(1-\frac{5}{3}(1-n)-a)\Gamma(1-\frac{1}{3}(1-n)-b)\Gamma(1-2(1-n)-c)}$$

$$\alpha_{g\circ f_3^3(z)}(k,n)=\frac{1}{k}(\frac{8}{3})^{n-k}\sum_{a,b,c|a+b+c=k-1}\frac{e^{(\frac{i\pi}{2}(a-b)-\frac{2i\pi c}{3})}2^{-c}}{a!b!c!}$$

$$\times\frac{\Gamma(1-\frac{5}{3}(1-n))\Gamma(1-\frac{1}{3}(1-n))\Gamma(1-2(1-n))}{\Gamma(1-\frac{5}{3}(1-n)-a)\Gamma(1-\frac{1}{3}(1-n)-b)\Gamma(1-2(1-n)-c)} \tag{5.122}$$

and

$$\beta_{I(z)}(k,l|n)=\frac{2}{(k+l-1)!}(-\frac{2}{3})^{n-k-l+1}\frac{\Gamma(-3)}{\Gamma(-1-k-l)}$$

$$\beta_{g\circ f_1^3(z)}(k,l|n)=\frac{1}{k+l-1}(-\frac{8}{3})^{n-k-l+1}$$

$$\times\sum_{a,b,c|a+b+c=k+l-2}\frac{e^{\frac{i\pi}{2}(a-b)}2^{-c}}{a!b!c!}$$

$$\times\frac{\Gamma(-\frac{7}{3})\Gamma(\frac{1}{3})\Gamma(-3)}{\Gamma(-\frac{7}{3}-a)\Gamma(\frac{1}{3}-b)\Gamma(-3-c)}$$

$$\beta_{g\circ f_2^3(z)}(k,l|n)=\frac{1}{k+l-1}(-\frac{8}{3})^{n-k-l+1}$$

$$\times\sum_{a,b,c|a+b+c=k+l-2}\frac{e^{(\frac{i\pi}{2}(a-b)+\frac{2i\pi c}{3})}2^{-c}}{a!b!c!}$$

$$\times\frac{\Gamma(-\frac{7}{3})\Gamma(\frac{1}{3})\Gamma(-3)}{\Gamma(-\frac{7}{3}-a)\Gamma(\frac{1}{3}-b)\Gamma(-3-c)}$$

$$\beta_{g\circ f_2^3(z)}(k,l|n)=\frac{1}{k+l-1}(\frac{8}{3})^{n-k-l+1}$$

$$\times\sum_{a,b,c|a+b+c=k+l-2}\frac{e^{(\frac{i\pi}{2}(a-b)-\frac{2i\pi c}{3})}2^{-c}}{a!b!c!}$$

$$\times\frac{\Gamma(-\frac{7}{3})\Gamma(\frac{1}{3})\Gamma(-3)}{\Gamma(-\frac{7}{3}-a)\Gamma(\frac{1}{3}-b)\Gamma(-3-c)} \tag{5.123}$$

Our notations are defined as follows. Consider a function of the normalized Bell polynomials $f(B_1,B_2,....)=\sum_{n>0}f_nB_{\vec{\psi}}^{(n)}$ and the associate function $f(B)$ given by the formal series in auxiliary argument $B$ $f(B)=\sum_{n>0}f_nB^n$ Consider a transformation $f(B)\to g(B)$ where $g(B)=\sum_n g_nB^n$ can obtained from $f$ by differentiation over $B$, integration, multiplication(s) by $B$ and/or their combination. Then the formal series for $g(B)$ define the new associate generating function of the normalized Bell polynomials $g(B_1, ..., B_n) = \sum_n g_n B_{\vec{\psi}}^{(n)}$ by identifying $B^n \to B_{\vec{\psi}}^{(n)}$.

This fully determines the transformations of the SFT string field ansatz under the conformal transformations mapping the worlsheets to the wedges of the single disc and then to the single half-plane. The next step is to point out the action of the BRST charge on the string field ansatz. This too can be reduced to the transformations of the ansatz functions $G$ and $H$. Since the only SFT correlator involving the BRST charge is $<< Q\Psi|\Psi >>= < Q\Psi(0) I \circ \Psi(0) >$ and both $\Psi$ and $I \circ \Psi$ are proportional to $c$, the only terms in the commutator with the BRST charge (1.9) contributing to this correlator are those proportional to $\partial cc$ and $\partial^2 cc$, while all the terms in $Q\Psi$ containing higher derivatives of the $c$-ghost don't contribute to the correlator since $\partial^n cc \sim B_\sigma^{(n-1)} \partial cc$ with $\sigma$ being the bosonized $c$-ghost. Such terms do not contribute to the two-point correlators since the Bell polynomials in the derivatives of the bosonized $c$-ghost: $B_\sigma^{(n-1)}$ cannot fully contract to the $c$-ghost of the opposite string field for $n > 2$. Using the operator algebra for the Bell polynomial operators, it is straightforward to show that, for our string field ansatz the relevant terms in the BRST transformation are given by:

$$
\begin{aligned}
& Q\Psi \equiv QcG(H(B)) = \partial cc(B\partial_B G(H(B)) - G(H(B))) \\
& + \frac{1}{2}\partial^2 ccB\partial_B^2 G(H(B)) \\
& = \partial c(B\partial_B \Psi - \Psi) + \frac{1}{2}\partial^2 cB\partial_B^2 \Psi
\end{aligned}
\tag{5.124}
$$

with the notations explained above. With all the above identities it is now straightforward to calculate the SFT correlators. The three-point correlator is then computed to give

$$
\begin{aligned}
& << g_1 \circ G(H(B))(0) g_2 \circ G(H(B))(0) g_3 \circ G(H(B))(0) >> \\
& \equiv << \Psi|\Psi \star \Psi >>= T_1 + T_2 + T_3 + T_4
\end{aligned}
\tag{5.125}
$$

Here

$$
\begin{aligned}
T_1 = & \sum_{n_1,n_2,n_3=0}^{\infty} \sum_{m_1,m_2,m_3=0}^{\infty} \sum_{q_1,q_2,q_3=0}^{\infty} \sum_{k_1=0}^{m_1} \sum_{k_2=0}^{m_2} \sum_{k_3=0}^{m_3} \sum_{N,R,T=0}^{\infty} \\
& \sum_{N|n_1...n_{q_1}} \sum_{R|r_1...r_{q_2}} \sum_{T|t_1...t_{q_3}} \sum_{N_1=0}^{N+m_1-k_1} \sum_{R_1=0}^{R+m_2-k_2} \sum_{T_1=0}^{T+m_3-k_3} \\
& \{ \frac{g_{q_1+1} g_{q_2+1} g_{q_3+1}}{q_1! q_2! q_3!} h_{m_1} h_{m_2} h_{m_3} h_{n_1} ....h_{n_{q_1}} h_{r_1} ...h_{r_{q_2}} h_{t_1} ....h_{t_{q_3}}
\end{aligned}
$$

$$\prod_{i_1=1}^{q_2+1}\prod_{j_1=1}^{q_1+1}\prod_{i_2=1}^{q_3+1}\prod_{j_2=1}^{q_1+1}\prod_{i_3=1}^{q_3+1}\prod_{j_3=1}^{q_2+1}\lambda_{\alpha_{i_1j_1}|\beta_{j_1i_1}}\lambda_{\tilde{\alpha}_{i_2j_2}|\beta_{j_2i_2}}\lambda_{\tilde{\alpha}_{i_2j_2}|\tilde{\beta}_{j_2i_2}}$$
$$\prod_{\mu=1}^{q_1+1}\sigma^{-1}(\lambda_{\alpha_{\mu,1}|\beta_{1,\mu}},...\lambda_{\alpha_{\mu,q_2+1}|\beta_{q_2+1,\mu}})$$
$$\times\sigma^{-1}(\lambda_{\tilde{\alpha}_{\mu,1}|\beta_{1,\mu}},...\lambda_{\tilde{\alpha}_{\mu,q_3+1}|\beta_{q_3+1,\mu}})$$
$$\prod_{\nu=1}^{q_2+1}\sigma^{-1}(\lambda_{\tilde{\alpha}_{\mu,1}|\tilde{\beta}_{1,\mu}},...\lambda_{\tilde{\alpha}_{\nu,q_3+1}|\tilde{\beta}_{q_3+1,\nu}})$$
$$\times\sigma^{-1}(s_1^{(1)},...s_{q_2+1}^{(1)})\sigma^{-1}(\tilde{s}_1^{(1)},...\tilde{s}_{q_3+1}^{(1)})\sigma^{-1}(s_1^{(2)},...s_{q_1+1}^{(2)})$$
$$\times\sigma^{-1}(\tilde{s}_1^{(2)},...\tilde{s}_{q_3+1}^{(2)})\sigma^{-1}(s_1^{(3)},...s_{q_1+1}^{(3)})$$
$$\sigma^{-1}(\tilde{s}_1^{(3)},...\tilde{s}_{q_2+1}^{(3)})((q_2+1)!(q_3+1)!)^{q_1+1}((q_3+1)!)^{q_2+1}$$
$$\times 2^{N+T_1-N_1+m_1-k_1}(\sqrt{3})^{N+R+T+m_1-k_1+m_2-k_2+m_3-k_3}\}$$
$$+permutations(g_1\circ\Psi,g_2\circ\Psi,g_3\circ\Psi) \tag{5.126}$$

with the exchange numbers defined similarly to the previous case, as well as the $\sigma^{-1}$-factors, defined by products of array multiplicities in the relevant partitions. Next, the $s,\tilde{s}$-numbers are similar to the $r,s$-numbers defined previously and are related to conformal dimension losses of string field components due to interactions with partition elements (individual Bell polynomials) in components of two opposite string fields. Altogether, these numbers satisfy the following constraints:

$$\sum_{j=1}^{q_2+1}\alpha_{ij}+\sum_{j=1}^{q_3+1}\tilde{\alpha}_{ij}=n_i(i=1,...,q_1+1)$$
$$\sum_{j=1}^{q_1+1}\beta_{ij}+\sum_{j=1}^{q_3+1}\tilde{\beta}_{ij}=r_i(i=1,...,q_2+1)$$
$$\sum_{j=1}^{q_1+1}\gamma_{ij}+\sum_{j=1}^{q_2+1}\tilde{\gamma}_{ij}=t_i(i=1,...,q_3+1) \tag{5.127}$$

and

$$\sum_{i=1}^{q_1+1}\alpha_{ij}=s_j^{(1)}(j=1,...,q_2+1)$$
$$\sum_{i=1}^{q_1+1}\tilde{\alpha}_{ij}=\tilde{s}_j^{(1)}(j=1,...,q_3+1)$$

$$\sum_{i=1}^{q_2+1} \alpha_{ij} = s_j^{(2)} (j = 1, ..., q_1 + 1)$$

$$\sum_{i=1}^{q_2+1} \tilde{\beta}_{ij} = \tilde{s}_j^{(2)} (j = 1, ..., q_3 + 1)$$

$$\sum_{i=1}^{q_3+1} \gamma_{ij} = s_j^{(3)} (j = 1, ..., q_1 + 1)$$

$$\sum_{i=1}^{q_2+1} \tilde{\gamma}_{ij} = \tilde{s}_j^{(3)} (j = 1, ..., q_2 + 1) \tag{5.128}$$

and furthermore

$$N + m_1 - k_1 = \sum_{i=1}^{q_1+1} n_i = \sum_{j=1}^{q_2+1} s_j^{(1)} + \sum_{j=1}^{q_3+1} \tilde{s}_j^{(1)}$$

$$R + m_2 - k_2 = \sum_{i=1}^{q_2+1} r_i = \sum_{j=1}^{q_1+1} s_j^{(2)} + \sum_{j=1}^{q_3+1} \tilde{s}_j^{(2)}$$

$$T + m_3 - k_3 = \sum_{i=1}^{q_3+1} t_i = \sum_{j=1}^{q_1+1} s_j^{(3)} + \sum_{j=1}^{q_2+1} \tilde{s}_j^{(3)} \tag{5.129}$$

In other words, the exchange numbers, that form the OPE structure of the Bell polynomial products, can be visualized as "partitions of partitions" of the conformal dimensions of the string field components.

This constitutes $T_1$, the first out of 4 terms contributing to the 3-point correlator. The remaining three can be obtained from $T_1$ by few simple replacements/manipulations. That is, $T_2$ is obtained from $T_1$ by replacing one of three $\alpha$ coefficients in (5.126) by the $\beta$-coefficient: $\alpha(k_1, n_1) \to \beta(k_1, l_1|n_1)$, with $k_1, l_1$ being summed over from 0 to $k_1 + l_1 = n_1$; inserting an extra h-coefficient in the sum according to: $h_{m_1} h_{m_2} h_{m_3} \to h_{m_1} h_{m_2} h_{m_3} h_{m_4}$, replacing the difference $m_1 - k_1 \to m_1 + m_4 - k_1 - l_1$ and finally replacing $q_1 + 1 \to q_1 + 2$ in the upper limits in the products over $j_1$ and $j_2$ in (5.127)-(5.129), as well as in the relevant $g$-coefficient (the first among three in (5.126)) and in the relevant $\sigma^{-1}$-factors, increasing their number of arguments by one unit - and finally, permuting over the conformal transformations by $g_1$, $g_2$, $g_3$, as in $T_1$. Thus the $T_1$-contribution has the $\alpha\alpha\alpha$-structure, while $T_2$ carries the $\beta\alpha\alpha$-structure. Similarly, to obtain $T_3$ out of $T_2$, one further replaces the second $\alpha$-coefficient by the $\beta$-coefficient: $\alpha(k_2, n_2) \to \beta(k_2, l_2|n_2)$, inserts an extra

$h$-coefficient in the product: $h_{m_1}...h_{m_4} \to h_{m_1}...h_{m_4}h_{m_5}$, and further replaces $m_2 - k_2 \to m_2 + m_5 - k_2 - l_2$ and $q_2 + 1 \to q_2 + 2$ according to the prescriptions explained above. This, upon the permutation over the conformal transformations, similar to the above, gives the $T_3$-contribution with the $\beta\beta\alpha$-structure. The final contribution, $T_4$, having the $\beta\beta\beta$-structure, is obtained similarly from $T_3$ by replacing the last remaining $\alpha$ with $\beta$ and performing the manipulations identical to those described above. The overall expression for the three-point correlator thus looks complex enough. Nevertheless, it is straightforward to check that, just as in the elementary warm-up example demonstrated previously, the complicated sum given by (5.130)-(5.134) can be converted successfully into the generating composite function and identified with its series expansion. Namely, we obtain:

$$
\begin{aligned}
&<< \Psi|\Psi \star \Psi >> \\
&\equiv << g_1 \circ G(H(B))(0) g_2 \circ G(H(B))(0) g_3 \circ G(H(B))(0) >> \\
&= \sum_{j=1}^{4} K_j(G(H(F_\lambda)))
\end{aligned}
\tag{5.130}
$$

where

$$
\begin{aligned}
&K_1(G(H(F_\lambda))) \\
&= \sum_{n_1=0}^{\infty}\sum_{n_2=0}^{\infty}\sum_{n_3=0}^{\infty}\sum_{k_1=0}^{n_1}\sum_{k_2=0}^{n_2}\sum_{k_3=0}^{n_3}\sum_{Q=1}^{k_1}\sum_{R=1}^{k_2}\sum_{S=1}^{k_3} \\
&\{[\alpha_{g_1}(k_1,n_1)\alpha_{g_2}(k_2,n_2)\alpha_{g_3}(k_3,n_3) \\
&+\alpha_{g_1}(k_1,n_1)\alpha_{g_3}(k_2,n_2)\alpha_{g_2}(k_3,n_3) \\
&+\alpha_{g_2}(k_1,n_1)\alpha_{g_1}(k_2,n_2)\alpha_{g_3}(k_3,n_3) \\
&+\alpha_{g_2}(k_1,n_1)\alpha_{g_3}(k_2,n_2)\alpha_{g_1}(k_3,n_3) \\
&+\alpha_{g_3}(k_1,n_1)\alpha_{g_2}(k_2,n_2)\alpha_{g_1}(k_3,n_3) \\
&+\alpha_{g_3}(k_1,n_1)\alpha_{g_1}(k_2,n_2)\alpha_{g_2}(k_3,n_3)\} \\
&\times\partial_{F_\lambda}^{k_1} H(F_\lambda(x,y))|_{x,y=\sqrt{3}}\partial_{F_\lambda}^{k_2} H(F_\lambda(x,y))|_{x,y=2\sqrt{3}} \\
&\times\partial_{F_\lambda}^{k_3} H(F_\lambda(x,y))|_{x,y=\sqrt{3}} \\
&\times G'(\partial^Q H(F_\lambda(x,y)))|_{x,y=\sqrt{3}} G'(\partial^R H(F_\lambda(x,y)))|_{x,y=2\sqrt{3}} \\
&\times G'(\partial^S H(F_\lambda(x,y)))|_{x,y=\sqrt{3}}
\end{aligned}
\tag{5.131}
$$

$$
\begin{aligned}
&K_2(G(H(F_\lambda))) = \\
&\sum_{n_1=0}^{\infty}\sum_{n_2=0}^{\infty}\sum_{n_3=0}^{\infty}\sum_{k_1,l_1=0}^{k_1+l_1=n_1}\sum_{k_2=0}^{n_2}\sum_{k_3=0}^{n_3}\sum_{Q=1}^{k_1+l_1-1}\sum_{R=1}^{k_2}\sum_{S=1}^{k_3} \\
&\beta_{g_1}(k_1,l_1|n_1)(\alpha_{g_2}(k_2,n_2)\alpha_{g_3}(k_3,n_3) \\
&+\alpha_{g_3}(k_2,n_2)\alpha_{g_2}(k_3,n_3)) \\
&\times G''(\partial^Q H(F_\lambda(x,y)))|_{x,y=\sqrt{3}}G'(\partial^R H(F_\lambda(x,y)))|_{x,y=2\sqrt{3}} \\
&\times G'(\partial^S H(F_\lambda(x,y)))|_{x,y=\sqrt{3}} \\
&\times(\partial_{F_\lambda}^{k_1} H(F_\lambda(x,y))|_{x,y=\sqrt{3}}\partial_{F_\lambda}^{l_1} H(F_\lambda(x,y))|_{x,y=\sqrt{3}} \\
&\times\partial_{F_\lambda}^{k_2} H(F_\lambda(x,y))|_{x,y=2\sqrt{3}}\partial_{F_\lambda}^{k_3} H(F_\lambda(x,y))|_{x,y=\sqrt{3}}) \\
&+\beta_{g_2}(k_1,l_1|n_1)(\alpha_{g_1}(k_2,n_2)\alpha_{g_3}(k_3,n_3) \\
&+\alpha_{g_3}(k_2,n_2)\alpha_{g_1}(k_3,n_3)) \\
&\times G''(\partial^Q H(F_\lambda(x,y)))|_{x,y=2\sqrt{3}}G'(\partial^R H(F_\lambda(x,y)))|_{x,y=\sqrt{3}} \\
&\times G'(\partial^S H(F_\lambda(x,y)))|_{x,y=\sqrt{3}} \\
&\times(\partial_{F_\lambda}^{k_1} H(F_\lambda(x,y))|_{x,y=2\sqrt{3}}\partial_{F_\lambda}^{l_1} H(F_\lambda(x,y))|_{x,y=2\sqrt{3}} \\
&\times\partial_{F_\lambda}^{k_2} H(F_\lambda(x,y))|_{x,y=\sqrt{3}}\partial_{F_\lambda}^{k_3} H(F_\lambda(x,y))|_{x,y=\sqrt{3}}) \\
&+\beta_{g_3}(k_1,l_1|n_1)(\alpha_{g_1}(k_2,n_2)\alpha_{g_2}(k_3,n_3) \\
&+\alpha_{g_2}(k_2,n_2)\alpha_{g_1}(k_3,n_3)) \\
&\times G'(\partial^Q H(F_\lambda(x,y)))|_{x,y=\sqrt{3}}G'(\partial^R H(F_\lambda(x,y)))|_{x,y=\sqrt{3}} \\
&\times G'(\partial^S H(F_\lambda(x,y)))|_{x,y=2\sqrt{3}} \\
&\times(\partial_{F_\lambda}^{k_1} H(F_\lambda(x,y))|_{x,y=\sqrt{3}}\partial_{F_\lambda}^{l_1} H(F_\lambda(x,y))|_{x,y=\sqrt{3}} \\
&\times\partial_{F_\lambda}^{k_2} H(F_\lambda(x,y))|_{x,y=\sqrt{3}}\partial_{F_\lambda}^{k_3} H(F_\lambda(x,y))|_{x,y=2\sqrt{3}})
\end{aligned}
\tag{5.132}
$$

$$
\begin{aligned}
&K_3(G(H(F_\lambda))) = \\
&\sum_{n_1=0}^{\infty}\sum_{n_2=0}^{\infty}\sum_{n_3=0}^{\infty}\sum_{k_1,l_1=0;k_1<l_1}^{k_1+l_1=n_1}\sum_{k_2,l_2=0;k_2<l_2}^{k_2+l_2=n_2} \\
&\sum_{k_3=0}^{n_3}\sum_{Q=1}^{k_1+l_1-1}\sum_{R=1}^{k_2+l_2-1}\sum_{S=1}^{k_3} \\
&(\beta_{g_1}(k_1,l_1|n_1)\beta_{g_2}(k_2,,l_2|n_2)+ \\
&\beta_{g_2}(k_1,l_1|n_1)\beta_{g_1}(k_2,,l_2|n_2))\alpha_{g_3}(k_3,n_3)
\end{aligned}
$$

$$
\begin{aligned}
&\times G''(\partial^Q H(F_\lambda(x,y)))|_{x,y=\sqrt{3}} G''(\partial^R H(F_\lambda(x,y)))|_{x,y=2\sqrt{3}}\\
&\times G'(\partial^S H(F_\lambda(x,y)))|_{x,y=\sqrt{3}}\\
&\times(\partial^{k_1}_{F_\lambda} H(F_\lambda(x,y))|_{x,y=\sqrt{3}} \partial^{l_1}_{F_\lambda} H(F_\lambda(x,y))|_{x,y=\sqrt{3}}\\
&\times \partial^{k_2}_{F_\lambda} H(F_\lambda(x,y))|_{x,y=2\sqrt{3}}\\
&\times \partial^{l_2}_{F_\lambda} H(F_\lambda(x,y))|_{x,y=2\sqrt{3}}) \partial^{k_3}_{F_\lambda} H(F_\lambda(x,y))|_{x,y=\sqrt{3}}\\
&+(\beta_{g_1}(k_1,l_1|n_1)\beta_{g_3}(k_2,,l_2|n_2)\\
&+\beta_{g_3}(k_1,l_1|n_1)\beta_{g_1}(k_2,,l_2|n_2))\alpha_{g_2}(k_3,n_3)\\
&\times G''(\partial^Q H(F_\lambda(x,y)))|_{x,y=2\sqrt{3}} G''(\partial^R H(F_\lambda(x,y)))|_{x,y=\sqrt{3}}\\
&\times G'(\partial^S H(F_\lambda(x,y)))|_{x,y=\sqrt{3}}\\
&\times(\partial^{k_1}_{F_\lambda} H(F_\lambda(x,y))|_{x,y=2\sqrt{3}} \partial^{l_1}_{F_\lambda} H(F_\lambda(x,y))|_{x,y=2\sqrt{3}}\\
&\times \partial^{k_2}_{F_\lambda} H(F_\lambda(x,y))|_{x,y=\sqrt{3}}\\
&\times \partial^{l_2}_{F_\lambda} H(F_\lambda(x,y))|_{x,y=\sqrt{3}}) \partial^{k_3}_{F_\lambda} H(F_\lambda(x,y))|_{x,y=\sqrt{3}}\\
&+(\beta_{g_2}(k_1,l_1|n_1)\beta_{g_3}(k_2,,l_2|n_2)\\
&+\beta_{g_3}(k_1,l_1|n_1)\beta_{g_2}(k_2,,l_2|n_2))\alpha_{g_1}(k_3,n_3)\\
&\times G''(\partial^Q H(F_\lambda(x,y)))|_{x,y=\sqrt{3}} G''(\partial^R H(F_\lambda(x,y)))|_{x,y=\sqrt{3}}\\
&\times G'(\partial^S H(F_\lambda(x,y)))|_{x,y=2\sqrt{3}}\\
&\times(\partial^{k_1}_{F_\lambda} H(F_\lambda(x,y))|_{x,y=\sqrt{3}} \partial^{l_1}_{F_\lambda} H(F_\lambda(x,y))|_{x,y=\sqrt{3}}\\
&\times \partial^{k_2}_{F_\lambda} H(F_\lambda(x,y))|_{x,y=\sqrt{3}}\\
&\times \partial^{l_2}_{F_\lambda} H(F_\lambda(x,y))|_{x,y=\sqrt{3}}) \partial^{k_3}_{F_\lambda} H(F_\lambda(x,y))|_{x,y=2\sqrt{3}}
\end{aligned}
\tag{5.133}
$$

$$
\begin{aligned}
&K_4(G(H(F_\lambda)))\\
&=\sum_{n_1=0}^{\infty}\sum_{n_2=0}^{\infty}\sum_{n_3=0}^{\infty}\sum_{k_1,l_1=0;k_1<l_1}^{k_1+l_1=n_1}\sum_{k_2,l_2=0;k_2<l_2}^{k_2+l_2=n_2}\\
&\sum_{k_3,l_3=0;k_3<l_3}^{n_3}\sum_{Q=1}^{k_1+l_1-1}\sum_{R=1}^{k_2+l_2-1}\sum_{S=1}^{k_3+l_3-1}\\
&(\beta_{g_1}(k_1,l_1|n_1)\beta_{g_2}(k_2,l_2|n_2)\beta_{g_3}(k_3,l_3|n_3)\\
&+\beta_{g_1}(k_1,l_1|n_1)\beta_{g_3}(k_2,l_2|n_2)\beta_{g_2}(k_3,l_3|n_3)\\
&+\beta_{g_1}(k_2,l_2|n_2)\beta_{g_2}(k_1,l_1|n_1)\beta_{g_3}(k_3,l_3|n_3)\\
&+\beta_{g_1}(k_2,l_2|n_2)\beta_{g_3}(k_1,l_1|n_1)\beta_{g_2}(k_3,l_3|n_3)
\end{aligned}
$$

$$\begin{aligned}
&+\beta_{g_1}(k_3,l_3|n_3)\beta_{g_2}(k_1,l_1|n_1)\beta_{g_3}(k_2,l_2|n_2)\\
&+\beta_{g_1}(k_3,l_3|n_3)\beta_{g_3}(k_1,l_1|n_1)\beta_{g_2}(k_2,l_2|n_2))\\
&\times G''(\partial^Q H(F_\lambda(x,y)))|_{x,y=\sqrt{3}}G''(\partial^R H(F_\lambda(x,y)))|_{x,y=\sqrt{3}}\\
&\times G''(\partial^S H(F_\lambda(x,y)))|_{x,y=2\sqrt{3}}\\
&\times(\partial^{k_1}_{F_\lambda}H(F_\lambda(x,y))|_{x,y=\sqrt{3}}\partial^{l_1}_{F_\lambda}H(F_\lambda(x,y))|_{x,y=\sqrt{3}}\\
&\times\partial^{k_2}_{F_\lambda}H(F_\lambda(x,y))|_{x,y=\sqrt{3}}\\
&\times\partial^{l_2}_{F_\lambda}H(F_\lambda(x,y))|_{x,y=\sqrt{3}}\partial^{k_3}_{F_\lambda}H(F_\lambda(x,y))|_{x,y=2\sqrt{3}}\\
&\times\partial^{k_3}_{F_\lambda}H(F_\lambda(x,y))|_{x,y=2\sqrt{3}})
\end{aligned}\tag{5.134}$$

where $G(\partial^Q H)$ is obtained from $G(H)$ by replacing the argument $H \to \partial^Q H$ and $\partial^Q H \equiv \frac{\partial^Q H}{\partial F_\lambda{}^Q}$ and finally $G'(\partial^Q H) = \frac{\partial G}{\partial(\partial^Q H)}$. This concludes the computation of the three-point SFT correlator for our solution ansatz. The final step is to compute the kinetic term $<< Q\Psi(0)|I\circ\Psi(0) >>$ using the operator products and the identities derived above. Given the BRST transformation of $\Psi$, this correlator is determined by two contributions: one proportional to the ghost part $< c(z)\partial cc(w > |_{z=0;w\to\infty} = (z-w)^2$ another to $< c(z)\partial^2 cc(w) > |_{z=0;w\to\infty} = -2(z-w)$. Note that, since $c$, $\partial cc$ and $\partial^2 cc$ ghost fields have conformal dimensions $-1$, $-1$ and 0 respectively, and since the conformal transformation by $I(z)$ takes 0 to infinity, it is straightforward to check that the matter part of the first contribution only contains the terms with the conformal dimensions of the string field components at $z$ equal to those of the string components at $w$; all the terms with unequal conformal dimensions of operators at $z$ and $w$ vanish in the limit $w \to \infty$. Similarly, the matter part of the second contribution (multiplied by the $< c(z)\partial^2 cc(w) > |_{z=0;w\to\infty}$ ghost correlator) only contains the terms with the conformal dimensions of the operators at $z$ equal to those of the operators at $w$ plus one.

Then, performing straightforward calculation of the correlator, similar to those above, plugging into SFT equations of motion leads to the defining relation for the $G(H(F_\lambda))$ function of our ansatz, given by:

$$\begin{aligned}
&\sum_{n=0}^{\infty}\sum_{k=1}^{n}\alpha_I(k,n)\sum_{Q=0}^{k}\{Z_0[\tilde{G}(H(F_\lambda))|G'(\partial^Q_{F_\lambda}H(F_\lambda))]\\
&-Z_1[\partial_{F_\lambda}\tilde{G}(H(F_\lambda))|G'(\partial^Q_{F_\lambda}\tilde{H}(F_\lambda))]\}]\partial^k_{F_\lambda}\tilde{H}(F_\lambda)|_{x=y=1}
\end{aligned}$$

$$
\begin{aligned}
&+\sum_{n=0}^{\infty}\sum_{k,l=1;k<l}^{n}\beta_I(k,l|n)\sum_{Q=0}^{k+l-1}\}Z_0[\tilde{G}(\tilde{H}(F_\lambda))|G''(\partial^Q_{F_\lambda}\tilde{H}(F_\lambda))]\\
&-Z_1[\partial_{F_\lambda}\tilde{G}(H(F_\lambda))|G''(\partial^Q_{F_\lambda}\tilde{H}(F_\lambda))]\}\\
&\times\partial^k_{F_\lambda}\tilde{H}(F_\lambda)|\partial^l_{F_\lambda}\tilde{H}(F_\lambda)|_{x,y=1}\}\\
&=\sum_j K_j(G(H(F_\lambda)))
\end{aligned}
\tag{5.135}
$$

where the operations $Z_k[f_1(x)|f_2(x)]$ acting on functions $f_1$ and $f_2$ are defined as follows: if $f_1(x) = \sum_m a_m x^m$ and $f_2(x) = \sum_n a_n x^n$ are the series expansions for $f_1$ and $f_2$ then $Z_k$ maps them into the function (formal series)

$$
Z_k[f_1(x)|f_2(y)] = \sum_n a_n b_{n+k} x^n y^{n+k} \tag{5.136}
$$

The tilde operations are again defined according to:

$$
\begin{aligned}
\tilde{G}(H(x)) &= x\frac{d}{dx}G(H(x)) - G(H(x))\\
\tilde{H}(x) &= x\frac{d}{dx}H(x) - H(x)
\end{aligned}
\tag{5.137}
$$

and the $K_j$-functions are defined by (5.130), (5.131). The functional equation (5.135) is our main result that constitutes the defining relation for the SFT ansatz. As cumbersome as this relation is, it can be, e.g., solved order by order by iterations and reduces the SFT equation of motionm to the identity which is essentially algebraic. In the case of $\alpha^2 = -2$ that we mostly have explored above, one particularly simple example for the generating functions solving the defining relation (5.135) is given by

$$
\begin{aligned}
G(H) &= \frac{1}{1-H(F_\lambda)}\\
H(x) &= \frac{1}{1-x}
\end{aligned}
\tag{5.138}
$$

Replacing $x^n \to B^{(n)}_{\vec{\psi}}$ according to our usual prescription, leads to the generating function of the SU(2) enveloping algebra $T(\mu)$ with the parameter $\mu$ defined by the $F_\lambda$ (elementary correlator of Bell polynomial operators). The classical $w_\infty$ algebra is then recovered in the simple pole $\sim (z-w)^{-1}$ of the OPE of $G(H(B^{(n)}(z)))$ and $G(H(B^{(n)}(w)))$. In general, the defining relation (5.135) parametrizes the class of SFT solutions, of which (5.138) is an elementary example. Finding the explicit form of this class of the solutions, generalizing (5.138) is an important challenge and obviously doesn't seem to be easy. However, it appears that this class is most naturally expressible in terms of the series in the powers of the generating function for the products of the Bell polynomial operators, relating it to the enveloping of the enveloping of $SU(2)$ (and more particularly, to the enveloping of $w_\infty$). The objects like these are known to be relevant to the quantization of higher spin theories and to the multi-particle realizations of the higher-spin algebras [83]. The crucial point about the SFT solutions, constrained to the subspace of operators given by products of the Bell polynomials, is that these objects

a) behave in a controllable and consistent way in the SFT star product computations

b) form a natural operator basis for the free-field realization of the $SU(2)$ envelopings and $w_\infty$.

So far we have considered the ansatz solution in bosonic string field theory, given by formal series in partial (incomplete) Bell polynomials of Bell polynomial operators in the worldsheet derivatives of the target space fields. These objects form an operator algebra realization for the enveloping of $SU(2)$, including the $w_\infty$ algebra appearing at the simple pole of the OPE. This, up to ideal factorization, is isomorphic to a chiral copy of higher spin algebra in $AdS_3$. The solution is given in terms of of the functional constraints on the generating functions for the operators realizing this enveloping. These constraints altogether are quite cumbersome and finding their manifest solutions doesn't appear to be an easy challenge, except for a relatively simple example (5.138).

Nevertheless, the constraints for the $h$ and $g$-expansion coefficients are essentially algebraic and in principle be analyzed order by order by iterations.

An important question is whether the construction, considered above, can be extended to SFT solutions involving higher-dimensional enveloping/higher spin algebras. A possible answer to that may come from superstring generalization of the computation performed above and switching on the $\beta - \gamma$ system of the superconformal ghosts. Just as the solution, considered above, was in a sense inspired by the bosonic $c = 1$ model (an elementary example pointing out the relevance of operator algebra involving Bell polynomial products to $w_\infty$ and higher spin algebra) one can use a supersymmetric $c = 1$ model coupled to the $\beta - \gamma$ system as a toy model inspiration. It is known that the interaction with the superconformal ghosts enhances the $SU(2)$ symmetry at the selfdual point to $SU(N)$ where $N-2$ is the maximal superconformal ghost number (ghost cohomology rank) of the generators. One can hope that manifest form of the vertex operators in this model would prompt us the form of the ansatz we should be looking for, and the resulting solution would be relevant to envelopings of $SU(N)$ or their subalgebras, related to isometries of $AdS$ in different space-time dimensions. The manifest form of the vertex operators in this model would again involve the products of Bell polynomials, however their structure will be far more diverse. So far, $\psi = \vec{\alpha}\vec{X}$ parameter of the operators was fixed to be the same for the all the string field components (as it is the same for all the operators for the bosonic discrete states and is equal to $-i\sqrt{2}X$). Switching on the higher superconformal ghost pictures in the $c = 1$ model would then result in the appearance of Bell polynomial products with mixed $\psi$-parameters. While the naive number of the parameters would be $\frac{1}{2}N(N-1)$ (total number of the lowering operators of $SU(N)$), the actual number would be less and of the order of $N$, since not all the lowering operators, acting on tachyonic primaries, lead to physically distinct states. The distinct states are basically generated by the lowering operators of ghost numbers $N-2$ carrying the maximum momentum value in the $X$-direction, equal to $N-1$, and the total number of such generators is $N-1$. Thus one can hope that introducing extra $\psi$-parameters will direct us towards the SFT solutions describing the higher-dimensional enveloping/ higher spin algebras. It looks plausible that the framework involving the WZW-type Berkovits string field theory may turn out to be a convenient framework for this program along with cubic superstring field theory with picture-changing insertions. Following this strategy, one can hope to find the defining constraints for the generating functions, similar to those considered above. It would certainly be of interest and of importance to study these constraints and to identify some of their manifest solutions. This hopefully shall lead to new important insights regarding nonperturbative higher spin configurations, as well as to deeper understanding of the underlying relations between SFT and higher spin field theories, which appear to be crucial ingredients of holography principle in general.

# Chapter 6

# Higher Spins as Rolling Tachyons in Open String Field Theory

## 6.1 Irregular Vertex Operators as Holographic Duals for Mixed-Symmetry Higher Spins

We already noticed on many occasions that higher spin fields in AdS space-time are the crucial ingredient of $AdS/CFT$ correspondence, as most of the composite operators on the conformal field theory (CFT) side are holographically dual to the higher spin modes. Let us now take a closer look on the structures of these composite operators. Generically, any operator on the CFT side carrying multiple space-time tensor indices, is expected to be dual to some higher-spin field with mixed symmetry. On the other hand, the higher spin holography implies that any correlator in boundary CFT is reproduced by the worldsheet correlators of vertex operators in string theory, i.e. any gauge-invariant observable on the CFT side has its dual vertex operator in AdS string theory. For simple some simple CFT/gauge theory observables such a correspondence is straightforward. For example, the string counterpart of $Tr(F^2)$ in super Yang-Mills theory is the vertex operator $V_\phi$ of a dilaton in string theory, while the stress-energy tensor $T_{mn}$ corresponds to the graviton's vertex operator $V_{mn}$, polarized along the $AdS$ boundary. One problem with checking this conjecture on the string theory side is that we know little about $AdS$ string dynamics beyond semiclassical approximation. This is related to the fact that the first-quantized string theory is background-dependent.

In addition, the higher spin interactions (at least beyond the quartic order) are known to be highly nonlocal, while the standard low-energy effective actions, stemming from vanishing $\beta$-function constraints in the first-quantized theories are typically local. This altogether suggests that the second-quantized formalism of the string field theory may be a more adequate formalism to approach the higher spin holography from the string theory side, especially given the formal similarity of background-independent string field theory (SFT) equations of motion and Vasiliev's equations in the unfolding formalism. This naturally poses a question of how the vertex operator to CFT observable correspondence may be extended off-shell, in particular involving the higher spin modes. This does not seem to be obvious. For example, consider a composite operator given by the $N$'th power of the stress-energy tensor in CFT:

$$T^n \sim T_{m_1 n_1} ... T_{m_N n_N} \tag{6.1}$$

In the gravity limit, this operator must be a dual of a certain field of spin $2N$ in $AdS$ with mixed symmetries. But what is the vertex operator description of such an object in string theory? To answer this question, one has to take the colliding limit of $N$ graviton vertex operators in string theory. Taking such a limit does not look simple and must not be confused with the normal ordering. Instead, in order to reproduce the correlation functions correctly in such a limit, one has to retain *all* the terms, up to *all* orders of the operator product expansion (OPE), as the operators are colliding at the common point. Such a limit is well-known in the matrix model formulations of Liouville and Toda theories and plays an important role in extending the $AGT$ conjecture to Argyres-Douglas type of supersymmetric gauge theories with asymptotic freedom. The result is given by rank $N-1$ irregular Gaiotto-BMT (Bonelli-Maruyoshi-Tanzini) states [132–135,137,140,141] These states extend the context of the primary operators in CFT and lead to special representations of Virasoro algebra, being the simultaneous eigenstates of $N+1$ Virasoro generators:

$$\begin{aligned} &L_n|U_N>=\rho_n|U_N> (N \leq n \leq 2N) \\ &L_n|U_N>=0(n>2N) \end{aligned} \tag{6.2}$$

It is straightforward to show that the irregular states admit the following irregular vertex operator representation in terms of Liouville or Toda fields:

$$\begin{aligned} &|U_N>=U_N|0> \\ &U_N=e^{\vec{\alpha_0}\vec{\phi}+\sum_{k=1}^N \vec{\alpha}_k \partial^k \vec{\phi}} \end{aligned} \tag{6.3}$$

where $\vec{\phi} = \phi_1, ...\phi_D$ is either $D$-component Toda field or parametrize the coordinates of $D$-dimensional target space in bosonic string theory. The $\vec{\alpha}_k$ parameters are related to the Virasoro eigenvalues (6.2) according to [137]

$$\rho_n = -\frac{1}{2} \sum_{k_1,k_2;k_1+k_2=n} \vec{\alpha}_{k_1}\vec{\alpha}_{k_2} \tag{6.4}$$

In case of $D \geq 2$, the irregular states, apart from being eigenvalues of positive Virasoro generators, are also the eigenstates of positive modes $W_n^{(p)}$ of the $W_n$-algebra currents ($3 \leq n \leq D+1$) where

$$W_n^{(p)} = \oint \frac{dz}{2i\pi} z^{p+n-1} W_n(z) \tag{6.5}$$

where $W_n$ are the spin $n$ primaries and $(n-1)N \leq p \leq nN$. so that

$$W_n^{(p)} U_N = \rho_n^p U_N \tag{6.6}$$

and $\rho_n^p$ are degree $n$ polynomials in the components of $\vec{\alpha}$. Note that, while the maximal possible rank $n$ is always at least $D+1$, for higher dimensions ($D > 5$) it is also possible to have the higher ranks $n > D+1$ as well. In general case, the upper bound on $n$ is in fact related to a rather complex problem in the partition theory. Namely, the maximal rank is given by the maximal number $n_{max}$ for which the inequality

$$\begin{aligned} &\sum_{k=1}^{n_{max}} \frac{(k+D-1)!\kappa(n_{max}|k)}{k!} \\ &- \sum_{q=1}^{n_{max}-1} \sum_{k=1}^{q} \frac{(k+D-1)!\kappa(n_{max}|k)}{k!} - (D-1)! \geq 0 \end{aligned} \tag{6.7}$$

where $\kappa(n|k)$ is the number of ordered partitions of $n$ with the length $k$: $n = p_1 + .... + p_k; 0 < p_1 \leq p_2 ... \leq p_k$.

The objects (6.3) are obviously not in the BRST cohomology and are off-shell (except for the regular case $\vec{\alpha}_k = 0; k \neq 0$ but make a complete sense in background-independent open string field theory. On the other hand, the $U_N$-vertices are related to the onshell vertex operators for the higher-spin fields. That is, $U_N$ is the generating vertex for the higher-spin operators through

$$V_{h.s.} = \sum_{s,\{k_1,...,k_s\}} H^{\mu_1...\mu_s}(\vec{\alpha}_0) \frac{\partial^s (cU_N)}{\partial\alpha_{k_1}^{\mu_1}...\partial\alpha_{k_s}^{\mu_s}} |_{\vec{\alpha}_k=0;k\neq 0} \tag{6.8}$$

where $c$ is the $c$-ghost, $H^{\mu_1 \dots \mu_s}(\vec{\alpha}_0)$ are the higher spin $s$ fields in the target space with masses $m = \sqrt{2(k_1 + \dots + k_s - 1)}$, at the momentum $\vec{\alpha}_0$ with all the due on-shell constraints on $H$ to ensure the BRST-invariance. Thus the correlation functions (irregular conformal blocks) of the $U_N$-vertices particularly encode the information about the higher-spin interactions in string theory. At nonzero $\vec{\alpha}_k$ the $U_N$ vertices generate the off-shell extensions of higher-spin wavefunctions, which can be studied using the string field theory techniques. A question of particular interest is to find the higher spin wavefunction configurations in terms of irregular vertex operators, solving SFT equations of motion analytically. In case if all $\vec{\alpha}_k = 0$, except for $\vec{\alpha}_0$, the $U_N$ vertex becomes a tachyonic primary. Then, multiplied by the space-time tachyon's wavefunction $T(\vec{\alpha}_0)$ $\Psi = c \int d^D \alpha_0 T(\vec{\alpha}_0) U_N$ is an elementary solution of of string field theory equations: $Q\Psi + \Psi \star \Psi = 0$ provided that $T$ satisfies the vanishing tachyon's $\beta$-function constraints $\beta_T = (\frac{1}{2}\alpha_0^2 - 1)T + const \times T^2 = 0$ in the leading order of string perturbation theory. This solution is elementary as it describes the *perturbative* background change by a tachyon. In the case of $\vec{\alpha}_k \neq 0$ things become far more interesting. The wavefunction in the string field $\Psi = c \int d^D \vec{\alpha}_0 \dots d^D \vec{\alpha}_N T(\vec{\alpha}_0, \dots, \vec{\alpha}_N) U_N$ can now be regarded as a generating wavefunction for higher spin excitations in string field theory with the SFT solution constraints on $T$ now related to nonperturbative background change due to higher spin excitations and the effective action on $T$ holding the keys to higher spin interactions at all orders, just like the well-known Schnabl's analytic solution [94] describes the physics around the minimum of nonperturbative tachyon potential (that would be calculated up to all orders, from the string perturbation theory point of view). Below we shall particularly concentrate on the rank 1 and search for the SFT analytic solutions in the form:

$$\Psi = c \int d^D \vec{\alpha} \int d^D \vec{\beta} T(\vec{\alpha}, \vec{\beta}) e^{\vec{\alpha}\vec{\phi} + \vec{\beta}\partial\vec{\phi}} \tag{6.9}$$

We find that, in the leading order, $\Psi$ is an analytic solution if $T$ satisfies the constraints, given by equations of motion described by the nonlocal effective action for generalized rolling tachyons. The nonlocality structures are controlled by the SFT worldsheet correlators and by the conformal transformations of the irregular blocks. The solution in particular provides a nice example of how the star product in string field theory translates into the Moyal product in the analytic SFT solutions. Although we explicitly concentrate on rank one case, the same structure appears to persist for higher irregular ranks as well. The effective equations of motion for $T(\vec{\alpha}, \vec{\beta})$

constitute the nonperturbative generalization of $\beta$-function equations in perturbative string theory, and possibly can be related to nonlocal field redefinitions for Vasiliev's invariant functionals [126].

## 6.2 Irregular Vertices as String Field Theory Solutions: Rank 1 Case

We start with the transformation properties of the irregular vertices under the conformal transformations $z \to f(z)$, necessary to compute the correlators in string field theory. Straightforward application of the stress tensor to the rank one irregular vertex gives infinitezimal conformal transformation:

$$\delta_\epsilon U_1(\vec{\alpha},\vec{\beta}) = [\oint \frac{dw}{2i\pi}\epsilon(w)T(w); U_1(\vec{\alpha},\vec{\beta},z)]$$
$$= \{\frac{1}{12}\partial^3\epsilon\beta^2 + \frac{1}{2}\partial^2\epsilon(\vec{\alpha}\vec{\beta}) + \partial\epsilon(\frac{1}{2}\alpha^2 + \vec{\beta}\frac{\partial}{\partial\vec{\beta}}) + \epsilon\partial_z\}U_1(\vec{\alpha},\vec{\beta},z) \tag{6.10}$$

It is not difficult to obtain the finite transformations for $U_1$, by integrating the infinitezimal transformations (6.10):

$$U_1(\vec{\alpha},\vec{\beta},z) = e^{\vec{\alpha}\vec{\phi}+\vec{\beta}\partial\vec{\phi}}$$
$$\to (\frac{df}{dz})^{\frac{\alpha^2}{2}} e^{\vec{\alpha}\vec{\phi}+\frac{df}{dz}\vec{\beta}\partial\vec{\phi}+(\vec{\alpha}\vec{\beta})\frac{d}{dz}log(\frac{df}{dz})+\frac{1}{12}S(f;z)} \tag{6.11}$$

where $S(f;z)$ is the Schwarzian derivative.

It is straightforward to generalize this result to transformation laws for the irregular vertices of arbitrary ranks. For the arbitrary rank N the BRST and finite conformal transformations for the irregular vertices have the form:

$$\{Q, cU_N\} = \{\oint \frac{dz}{2i\pi}(cT - bc\partial c); ce^{i\sum_{q=1}^{N}\vec{\alpha_q}\partial^q\vec{\phi}}\}$$
$$= \frac{1}{2}\sum_{q_1=0}^{N}\sum_{q_2=0}^{N}\frac{q_1!q_2!}{(q_1+q_2+1)!}(\vec{\alpha_{q_1}}\vec{\alpha_{q_2}}) : \partial^{q_1+q_2+1}ccU_N$$
$$+i\sum_{q=1}^{N}\sum_{p=1}^{q-1}\frac{q!}{p!(q-p)!}\partial^{q-p}cc(\vec{\alpha}_q\frac{\partial}{\partial\vec{\alpha}_{p+1}})U_N \tag{6.12}$$

and

$$
\begin{aligned}
&U_N \to (\frac{df}{dz})^{\frac{\alpha_0^2}{2}} e^{-\sum_{q_1,q_2=1}^{N} S_{q_1|q_2}(f;z)} \\
&e^{i\sum_{q=2}^{N}\sum_{k=1}^{q-1}\sum_{l=1}^{k} \frac{(q-1)!}{k!(q-1-k)!}\frac{d^{q-k}f}{dz^{q-k}} B_{k|l}(\partial f ... \partial^{k-l+1} f)(\vec{\alpha}_q \partial^{l+1}\vec{\phi})} \\
&\times e^{\partial^n f(\vec{\alpha}_q \partial\vec{\phi})}
\end{aligned}
\tag{6.13}
$$

where $S_{q_1|q_2}(f;z)$ are the generalized Schwarzian derivatives of the rank $(q_1, q_2)$, which explicit form is given in the section 7.2.

The evaluation of the kinetic term with the string field of the form (6.9) leads to

$$
\begin{aligned}
&<< Q\Psi(0) \star \Psi(0) >>=< Q\Psi(0) I \circ \Psi(0) > \\
&= lim_{w\to\infty} \int d^D\alpha_1 \int d^D\beta_1 \int d^D\alpha_2 \\
&\int d^D\beta_2 T(\vec{\alpha}_1, \vec{\beta}_1) y T(\vec{\alpha}_2, \vec{\beta}_2) w^{\alpha_2^2-1} e^{-\frac{\vec{\alpha}_2\vec{\beta}_2}{z}} \\
&\times[(\frac{1}{2}\alpha_1^2 - 1 + \vec{\beta}\frac{\partial}{\partial\vec{\beta}}) < \partial c c e^{i\vec{\alpha}_1\vec{\phi}+i\vec{\beta}_1\partial\vec{\phi}}(0) c e^{i\vec{\alpha}_1\vec{\phi}+iw^2\vec{\beta}_1\partial\vec{\phi}}(w) > \\
&+\vec{\alpha}_1\vec{\beta}_1 < \partial^2 c c e^{i\vec{\alpha}_1\vec{\phi}+i\vec{\beta}_1\partial\vec{\phi}}(0) c e^{i\vec{\alpha}_1\vec{\phi}+iw^2\vec{\beta}_1\partial\vec{\phi}}(w) > \\
&+\frac{\beta^2}{12} < \partial^3 c c e^{i\vec{\alpha}_1\vec{\phi}+i\vec{\beta}_1\partial\vec{\phi}}(0) c e^{i\vec{\alpha}_1\vec{\phi}+iw^2\vec{\beta}_1\partial\vec{\phi}}(w) >]
\end{aligned}
\tag{6.14}
$$

First of all, this correlator is only well-defined in case if the constraint

$$
\vec{\alpha}\vec{\beta} = 0 \tag{6.15}
$$

is imposed. Since for regular vertex operators $\alpha$ has a meaning of the momentum, the orthogonality constraint particularly implies that the $\beta$ parameter may be related to the Fourier image of the extra coordinates in space-time in the context of double field theory and $T$-duality (see the discussion section).

Furthermore, note that since the ghost correlator $< \partial^n cc(z_1)c(z_2) >= 0$ for $n > 2$, combined with the orthogonality constraint the only surviving terms in the correlator are those proportional to $\sim \partial cc$ in $Q\Psi$. In addition, in the on-shell limit $\alpha_0^2 \to 2$ the correlators involving the terms $\sim \partial cc\vec{\beta}\frac{\partial}{\partial\vec{\beta}}$ and $\sim \partial^2 cc$ are of the order of $\sim \frac{1}{w}$ and vanish.

Thus the only contributing correlator in the kinetic term gives

$$<< Q\Psi(0) \star \Psi(0) >>= lim_{w\to\infty} \int d^D\alpha_1 d^D\beta_1$$

$$\int d^D\alpha_2 d^D\beta_2 T(\vec{\alpha}_1,\vec{\beta}_1)T(\vec{\alpha}_2,\vec{\beta}_2)w^{\alpha_2^2-1}e^{-\frac{\vec{\alpha_2}\vec{\beta_2}}{z}}$$

$$\times\{(\frac{1}{2}\alpha_1^2 - 1 + \vec{\beta}\frac{\partial}{\partial\vec{\beta}}) < \partial cce^{i\vec{\alpha}_1\vec{\phi}+i\vec{\beta}_1\partial\vec{\phi}}(0)ce^{i\vec{\alpha}_1\vec{\phi}+iw^2\vec{\beta}_1\partial\vec{\phi}}(w) >\}$$

$$= \int d^D\alpha d^D\beta\frac{1}{2}(\alpha^2-1)e^{\beta^2}T(\vec{\alpha},\vec{\beta})T(-\vec{\alpha},-\vec{\beta}) \tag{6.16}$$

where we used the orthogonality condition. This concludes the computation of the rank 1 contribution to the kinetic term in the SFT equations of motion. Note that, in the regularity limit $\beta^2 \to 0$ (coinciding with the on-shell limit for the rank 1 irregular operator), one can expand the exponent so that the kinetic term in the Lagrangian becomes

$$\sim \int d\alpha d\beta T(-\vec{\alpha},-\vec{\beta})(\frac{1}{2}\alpha^2+\beta^2-1)T(\vec{\alpha},\vec{\beta})+...$$

$$\sim \int dxdyT(x,y)(\frac{1}{2}\Box_x+\Box_y+1)T(x,y)+... \tag{6.17}$$

where we skipped the higher derivative terms. The next step is to calculate the cubic terms in the SFT equations. We have:

$$<< \Psi\star\Psi\star\Psi >>$$

$$= \int\prod_{j=1}^{3} d\alpha_j d\beta_j T(\vec{\alpha}_j,\vec{\beta}_j) < g_j^3 \circ cqe^{i\vec{\alpha}_j\vec{\phi}+i\vec{\beta}_j\partial\vec{\phi}}(0) > \tag{6.18}$$

Evaluating the values of $g_j^3$ and their Schwarzian derivatives at 0 and substituting the transformation laws for $\Psi$ under $g_j^3$, as well as the on-shell constraints on $\vec{\alpha}$, it is straightforward to calculate:

$$<< \Psi\star\Psi\star\Psi >>$$

$$= \int\prod_{j=1}^{3} d\alpha_j d\beta_j T(\vec{\alpha}_j,\vec{\beta}_j)$$

$$= e^{\frac{5}{54}(\beta_1^2+\beta_2^2+\beta_3^2)}(-\frac{2}{3})^{\frac{1}{2}\alpha_1^2-1}(-\frac{8}{3})^{\frac{1}{2}\alpha_2^2+\frac{1}{2}\alpha_3^2-2}$$

$$\times < e^{i\vec{\alpha_j}\vec{\phi}-\frac{2i}{3}\vec{\beta}_j\partial\vec{\phi}}(0)e^{i\vec{\alpha_j}\vec{\phi}-\frac{8i}{3}\vec{\beta}_j\partial\vec{\phi}}(\sqrt{3})e^{i\vec{\alpha_j}\vec{\phi}-\frac{8i}{3}\vec{\beta}_j\partial\vec{\phi}}(-\sqrt{3}) >>$$

$$= \int \prod_{j=1}^{3} d\alpha_j d\beta_j T(\vec{\alpha}_j, \vec{\beta}_j)$$
$$exp\{\frac{5}{54}(\beta_1^2+\beta_2^2+\beta_3^2)+\frac{16}{9}(\vec{\beta}_1\vec{\beta}_2+\vec{\beta}_1\vec{\beta}_3+\vec{\beta}_2\vec{\beta}_3)+\frac{4}{3\sqrt{3}}(\vec{\alpha}_2\vec{\beta}_3$$
$$-\vec{\alpha}_3\vec{\beta}_2)+\frac{2}{3\sqrt{3}}(4\vec{\alpha}_1(\vec{\beta}_3-\vec{\beta}_2)+\vec{\beta}_1(\vec{\alpha}_3-\vec{\alpha}_2))\}\delta(\sum_j \beta_j)\delta(\sum_j \alpha_j)$$
$$= \int \prod_{j=1}^{2} d\alpha_j d\beta_j T(\vec{\alpha}_1, \vec{\beta}_1)T(\vec{\alpha}_2, \vec{\beta}_2)T(-\vec{\alpha}_1-\vec{\alpha}_2, -\vec{\beta}_1-\vec{\beta}_2)$$
$$exp\{-\frac{43}{27}(\vec{\beta}_2\vec{\beta}_3+\beta_2^2+\beta_3^2)+\frac{2}{\sqrt{3}}(\vec{\alpha}_2\vec{\beta}_3-\vec{\alpha}_3\vec{\beta}_2)\} \tag{6.19}$$

Comparing the two-point and the three-point correlators, the irregular ansatz (6.9) solves the OSFT equation of motion provided that the wave-function $T(\vec{\alpha}, \vec{\beta})$ satisfy the Euler-Lagrange equation following from the cubic nonlocal effective action:

$$S = -\int d^D x d^D y \{T(x,y) e^{-\Box_y}(-\frac{1}{2}\Box_x - 1)T(x,y) + \tilde{\star}\{\tau^3(x,y)\} \tag{6.20}$$

where

$$\tau(x,y) = e^{-\frac{43}{27}\Box_y} T(x,y) \tag{6.21}$$

is a new (nonlocal) field variable, familiar from rolling tachyon cosmology and the star product with the tilde is defined according to

$$\tilde{\star}\{T_1(x,y)...T_N(x,y)\}$$
$$= lim_{y_1,\dots,y_N\to\{y} e^{\sum_{i,j=1;i<j}^{N} \frac{43}{27}\partial_{y_i}\partial_{y_j}} T(x,y_1)...T(x,y_N)\} \tag{6.22}$$

This defines the analytic open string field theory solution in terms of rank one irregular vertex operators, generating the higher-spin vertices on the leading Regge trajectory. The generating wavefunction for higher spins is thus described, in the leading order, by the nonlocal action (6.17). The actions of the type (6.17) are well known, as they describe extensions of rolling tachyon dynamics [49, 50], relevant to cosmological models with phantom fields. The nonlocality coefficients appearing in the analytic solution (1.9), (2.13) must be related to cosmological parameters of these models, such as dark energy state parameter and the vacuum expectation values of the rolling tachyon in the equilibrium limit (with the SFT solution interpolating between two vacua, describing the one dressed tachyon's value $\tau \sim e^{const\Box}T$ evolving into the vacuum state satisfying the Sen's conjecture constraints [145, 147]. The solution (6.9), (6.17) also defines the deformations of the BRST charge; solving the OSFT equations with the deformed charge would then result in quartic and higher order corrections in $\tau$. In the commutative level, nonlocal cosmological models of that type have been considered in a number of works (e.g. see [142–144, 148]). In the case $\vec{\beta} = 0$ (the regular case with the higher spins decoupled) the solution (6.9), (6.17) simplifies and is described by the local cubic action, which is just the leading order low-energy effective action for a tachyon in string perturbation theory. The solution (6.9), (6.17) is then the elementary one, describing the perturbative background deformation of flat target space in the leading order of the tachyon's $\beta$-function. With $\beta$-parameter switched on, the higher-spin dynamics enters the game and the effective action becomes non-local, describing *nonperturbative* background deformation in open string field theory. The rank one solution, considered so far, can be understood as the one describing generating wavefunction for higher-spin operators on the leading trajectory. It is then straightforward to extend this computation to describe the SFT solutions involving the irregular blocks of higher ranks, generating the higher-spin vertices on arbitrary Regge trajectories. The the effective action describing the generating higher-spin wavefunction essentially remains the same: in the leading order, it is cubic in $\tau = e^{\sum_j=1^q a_j \Box_{\beta_j}} T(\vec{\alpha}, \vec{\beta}_1, ...\vec{\beta}_q)$ ($a_j$ are the constants defining the OSFT solution) with the structure

$$\sim \int d\alpha \prod_j d\beta_j e^{\sum_j=1^q b_j \Box_{\beta_j}} T(\frac{1}{2}\alpha^2 - 1)T + \tilde{\star}(\tau^3) \tag{6.23}$$

where $\tau$ is again related to $T$ through nonlocal field redefinition.

All the family of the effective actions for collective higher-spin wavefunctions is essentially nonlocal. Clearly, they must be related to the nonlocalities and the star products appearing in higher-spin theories and Vasiliev's equations. It is also remarkable that the generating wavefunction for higher spin fields thus emerges in the context of the rolling tachyon cosmology. Since the solutions of the type (6.9), (6.17) generally describe the nonperturbative deformation of the flat background to collective higher-spin vacuum, it is a profound question whether such a deformation, related to cosmological evolution of generalized rolling tachyon type objects, is subject to constraints set up by the Sen's conjecture [145, 147].

## Chapter 7

# An Analytic Formula for Numbers of Restricted Partitions from Conformal Field Theory

The general structures for the higher-spin amplitudes, as well as for the string field theory solutions, describing the background deformations into higher-spin vacua, involve summations over patterns defined on the partitions of natural numbers, such as the correlation functions involving Bell polynomial operators. As we have seen, these operators are the important building blocks both for the analytic SFT solutions and for the vertex operators describing the higher-spin dynamics in AdS. This is is not incidental since the Bell polynomials are the natural objects appearing both in higher-spin algebras (as well as Vasiliev's free differential algebras). and are useful for classifications of irreducible classifications of higher-spin fields with mixed symmetries. For example, the number of terms in restricted Bell polynomial $B_{N|k}$ counts the total number of irreducible representations for mixed-symmetry higher spin $N$ with $k$ rows. This number is, in turn, given by the number $\lambda(N|k)$ of the ordered length $k$ partitions of number $N \geq k$. Counting numbers of such partitions is an important long-standing problem in the number theory, with plenty of possible applications, including the structures of modular forms. In this concluding section, we will how the exact solution for this problem can be described by using the methods of two-dimensional conformal field theory, in terms of the $2d$ correlators of irregular vertex operators, considered in the previous sections.

Let

$$
\begin{aligned}
&N = n_1 + n_2 + ... + n_k (1 \leq k \leq N) \\
&0 < n_1 \leq n_2 ... \leq n_k
\end{aligned}
\tag{7.1}
$$

be the length $k$ partition of a natural number $N$, $\lambda(N|k)$ be the number of such length $k$ partitions of $N$ and

$$\lambda(N) = \sum_{k=1}^{N} \lambda(N|k) \tag{7.2}$$

be the total number of partitions. Physically, $\lambda(N|k)$ and $\lambda(N)$ count the number of Young diagrams with $N$ cells and $k$ rows, and the total number $\lambda(N)$ of the diagrams with $N$ cells, and therefore are related to counting irreducible representations for higher-spin fields with spin value $N$. As it is well-known from number theory, obtaining exact analytic expressions for $\lambda(N)$ and especially for $\lambda(N|k)$ (say, in terms of some finite series) is a hard long-standing problem. For $\lambda(N)$, various asymptotic formulae are known for the large $N$ limit. The oldest and perhaps the best-known formula for $\lambda(N)$ was obtained by Ramanujan and Hardy in 1918 [108] and is given by:

$$\lambda(N) \sim \frac{1}{4N\sqrt{3}} e^{\pi\sqrt{\frac{2N}{3}}}$$

There are several improvements of this formula, notably by Rademacher [109], [110] who expressed $\lambda(N)$ in terms of infinite convergent series:

$$\lambda(N) = \frac{1}{\pi\sqrt{2}} \sum_{n=1}^{\infty} \sqrt{n}\alpha_n(N) \frac{d}{dN} \{ \frac{sinh[\frac{\pi}{n}\sqrt{\frac{2}{3}(N-\frac{1}{24})}]}{\sqrt{N-\frac{1}{24}}} \}$$

where

$$\alpha_n(N) = \sum_{0 \leq m \leq n; [m|n]} e^{i\pi(s(m,n) - \frac{2Nm}{n})}$$

with the notation $[m|n]$ implying the sum over $m$ taken over the values of $m$ relatively prime to $n$ and

$$s(m,n) = \frac{1}{4n} \sum_{k=1}^{n-1} cot(\frac{\pi k}{n}) cot(\frac{\pi k m}{n})$$

is the Dedekind sum for co-prime numbers.

The problem of finding $\lambda(N|k)$ is well-known to be even more tedious (see e.g. [121–123] for the discussion of Ramanujan-Rademacher type asymptotics for the restricted partitions). In this chapter, we will study the two-point short-distance correlator of irregular vertex operators in Conformal Field Theory [111, 116] that counts the number $\lambda(N|k)$ of restricted partitions, reproducing the well-known generating function for the partitions, when computed in the upper half-plane. One of these operators is the special case of rank one irregular vertex operators [112–115], that can be physically interpreted as a "dipole" in the Liouville theory (in the same sense that regular vertex operators, or primary fields, are the "charges"); another is related to a class of analytic solutions in open string field theory [117, 118], interpolating between flat and AdS backgrounds.

Next, we investigate the behaviour of this correlator under the peculiar class of conformal transformations that shrink the dipole's size to zero and reduce the correlator to contribution from zero modes of the irregular vertices. This leads to nontrivial identities involving the restricted partitions, expressing them in terms of generalized higher-derivative Schwarzians of these conformal transformations. In particular, this allows to express the number $\lambda(N|k)$ of the restricted partitions in terms of the finite series of the generalized Schwarzians and incomplete Bell polynomials of the conformal transformations considered, leading to the main result described in this chapter. Taking the short-distance limit in the correlation functions is necessary in order to be able to integrate the Ward identities, accounting for the local part of the two-dimensional conformal symmetry (or physically, the "spontaneous breaking" of the conformal symmetry for transformations with non-zero Schwarzians, i.e. other than fractional-linear). In general, such an integration is hard to perform and the correlators, computed in different coordinates (related by the transformation) differ by the infinite sum over Schwarzians and their higher-derivative counterparts. This difference, however, becomes controllable in the short-distance limit and, for the conformal transformations with the asymptotics considered below, can be compensated by a relatively simple factor, derived below.

## 7.1 Generating Function for Partitions: The Correlator

As it is well-known, $\lambda(N)$ and $\lambda(N|k)$ can be realized as expansion coefficients of the following (respectively) generating functions:

$$F(x) = \prod_{n=1}^{\infty} \frac{1}{1-x^n} = \sum_{N=0}^{\infty} \lambda(N) x^N$$

$$F(x,y) = \prod_{n=1}^{\infty} \frac{1}{1-yx^n} = \sum_{N=0,k=0;k\leq N}^{\infty} \lambda(N|k) x^N y^k$$

Unfortunately, these generating functions by themselves are not very helpful for elucidating explicit expressions for the partition numbers: taking their derivatives just gives trivial identities of the form $\lambda(N|k) = \lambda(N|k)$. For this reason, our strategy below will be to

1) identify the two-point correlators in Conformal Field Theory ($CFT$) counting the partitions (reproducing the generating function $F(x, y)$) in certain coordinates, namely, an upper half-plane;

2) using the conformal symmetry and suitable conformal transformations (identified below), derive the identities for the generating function, casting it in terms of an expression, making it possible to obtain an exact analytic expression $\lambda(N|k)$ in terms of the finite series. For simplicity, we shall concentrate on $c = 1$ CFT (free massless bosons in two dimensions). With some effort, it is straightforward to identify the two-point correlator, counting the partitions on the upper half-plane. This correlator is given by

$$G(\alpha, \beta, \epsilon) =< U_\alpha(z_1)V_\beta(z_2) > |_{z_1=i\epsilon;z_2=0}$$
$$\epsilon > 0 \tag{7.3}$$

where

$$U_\alpha(z) =: \prod_{n=0}^{\infty} \frac{1}{1 - \frac{\alpha^n \partial^n \phi}{n!}} : (z)$$
$$V_\beta(w) =: e^{\beta\partial\phi} : (w)$$

where the :: symbol stands for the normal ordering of operators in two-dimensional CFT, $\phi$ is $D = 2$ boson (e.g. a Liouville field, an open string's target space coordinate or a bosonized ghost), $\epsilon \to 0$, $\alpha$ and $\beta$ are the parameters that are introduced to control the generating function for the partitions. Both $U_\alpha$ and $V_\beta$ are understood in terms of formal series in $\alpha$ and $\beta$, with each term in the series being normally ordered by definition. To simplify notations , here and below we shall often use the partial derivative symbol for $z$-derivatives, even though in our case it coincides with ordinary derivative, since we only consider holomorphic sector.

Indeed, expanding in $\alpha$:

$$U_\alpha = \sum_{k=1}^{\infty} \sum_{n_1 \leq ... \leq n_k = 0}^{\infty} \sum_{p_1,...,p_k} \frac{\alpha^{p_1 n_1 + p_2 n_2 + ... + p_k n_k}}{(n_1!)^{p_1} ... (n_k!)^{p_k}}$$
$$\times : (\partial^{n_1}\phi)^{p_1} ... (\partial^{n_k}\phi)^{p_k} :$$

using the operator product expansion (OPE):

$$\partial^n \phi(z) : e^{\beta\partial\phi} : (w) \sim \frac{(-1)^{n+1} n! \beta : e^{\beta\partial\phi} : (w)}{(z-w)^{n+1}} + \ldots$$

and introducing $N = \sum p_k n_k$ one easily calculates

$$G(\alpha, \beta|\epsilon) = < U_\alpha(z) V_\beta(w) > |_{z=i\epsilon; w=0}$$

$$= \sum_{[N|n_1 \ldots n_k]=0}^{\infty} \sum_{k=0}^{N} \frac{\alpha^N \beta^k \lambda(N|k)}{(w-z)^{N+k}} = \prod_{n=1}^{\infty} \frac{1}{1 - \tilde{\alpha}^n \tilde{\beta}}$$

$$\tilde{\alpha} = \frac{\alpha}{w-z}; \tilde{\beta} = \frac{\beta}{w-z}$$

i.e. $G$ is the generating function for restricted partitions with

$$\lambda(N|k) = \frac{(-i\epsilon)^{N+k}}{N!k!} \partial_\alpha^N \partial_\beta^k G(\alpha, \beta|\epsilon)|_{\alpha,\beta=0}$$

Now that we have identified the correlator generating $\lambda(N|k)$, the next step is to identify the suitable conformal transformation. Note that the operator $V_\beta$ is the special case of rank one irregular vertex operator, creating a simultaneous eigenstate of Virasoro generators $L_1$ and $L_2$ (with eigenvalues 0 and $\frac{\beta^2}{2}$ respectively) and physically can be understood as a dipole with the size $\beta$. For this reason, it is natural to choose the transformation such that the dipole's size shrinks to zero in the new coordinates. So we will consider the conformal transformations of the form

$$z \to f(z) = h(z) e^{-\frac{i}{z}} \tag{7.4}$$

where $h(z)$ is regular and at 0, $h(0) \neq 0$ and it it is smooth and analytic in the upper half-plane (perhaps except for infinity) In particular, it is instructive to consider $h(z) = 1$ and $h(z) = cos(z)$. Now we have to:

1. Compute infinitezimal transformations of $U_\alpha$ and $V_\beta$.

2. Integrate them to get the finite transformations for $U_\alpha$ and $V_\beta$ under $f(z)$.

3. Since $f(z)$ is not a fractional-linear transformation, and its Schwarzian is singular at 0, to match the correlators in different coordinates, one has to take into account the "spontaneous symmetry breaking" of the conformal symmetry (with higher Virasoro modes playing the role of "Goldstone modes"), by integrating the Ward identities for $f(z)$ and regularizing the final expression, in order to ensure that the correlators computed in two coordinates match upon $f(z)$.

## 7.2 Partition-Counting Correlator: The Conformal Transformations

An important building block in our computation involves the finite conformal transformation laws for the operators of the form $T^{(n_1,n_2)} = \frac{1}{n_1!n_2!} : \partial^{n_1}\phi\partial^{n_2}\phi :$ of conformal dimensions $n_1 + n_2$ - in fact, the final answer for the number of partitions will be expessed in terms of the finmite series in the higher-derivative Schwarzians of $f(z) = e^{-\frac{i}{z}}$.

In case of $n_1 = n_2 = 1$ the operator $T^{(1,1)}$ (up to normalization constant of $\frac{1}{2}$) is just the stress-energy tensor, and both its infinitezimal and finite transformation laws are well-known.

The infinitezimal transformation of $T^{(1,1)}$ under $z \to z + \epsilon(z)$ is

$$\delta_\epsilon T^{(1,1)}(z) = \epsilon\partial T^{1,1}(z) + 2\partial\epsilon T^{(1,1)}(z) + \frac{1}{6}\partial^3\epsilon(z) \tag{7.5}$$

(here and everywhere below the infinitezimal conformal transformation parameter $\epsilon(u)$ is not to be confused with the $i\epsilon$ for the location of $z$ which hopefully will always be clear from the context; in our notations the former always will appear with the argument, while the latter will not). This infinitezimal transformation can be integrated to give the finite conformal transformation law for any $f(z)$ according to

$$T^{(1,1)}(z) \to (\frac{df(z)}{dz})^2 T^{(1,1)}(f(z)) + S_{1|1}(f;z) \tag{7.6}$$

where $S_{1|1}$ is (up to the conventional normalization factor of $\frac{1}{6}$) the Schwarzian derivative, defined according to:

$$S_{1|1}(f;z) = \frac{1}{6}(\frac{f''(z)}{f'(z)})' - \frac{1}{12}(\frac{f''(z)}{f'(z)})^2 \tag{7.7}$$

The integrated transformation (7.6) can be obtained from (7.5) e.g. by requiring that (7.5) is reproduced from (7.6) in the infinitezimal limit and that the composition of two transformations $f(z)$ and $g(z)$ gives again the conformal transformation with $f(g(z))$. Likewise, we can derive the transformation rules for an arbitrary $T^{(n_1,n_2)}$. The infinitezimal transformation is

$$\begin{aligned}\delta_\epsilon T^{(n_1,n_1)}(z) &= [\frac{1}{2}\oint \frac{du}{2i\pi}\epsilon(u) : (\partial\phi)^2 : (u); T^{(n_1,n_1)}(z)] \\ &= \frac{1}{n_2!} : \partial^{n_1}(\epsilon\partial\phi)\partial^{n_2}\phi : (z) + \frac{1}{n_1!} : \partial^{n_1}\phi\partial^{n_2}(\epsilon\partial\phi) : (z) \\ &+ \frac{1}{(n_1+n_2+1)!}\partial^{n_1+n_2+1}\epsilon(z)\end{aligned} \tag{7.8}$$

and can be integrated, by imposing the similar constraints:

$$T^{(n_1,n_2)}(z) \to \frac{1}{n_1!n_2!} : \delta_{n_1}(\phi; f(z))\delta_{n_2}(\phi; f(z)) : +S_{n_1|n_2}(f; z) \tag{7.9}$$

where the operators

$$\begin{aligned}
&\delta_n(\phi, f) = \frac{\partial^n f}{\partial z^n}\partial\phi(f(z)) \\
&+\sum_{k=1}^{n-1}\sum_{l=1}^{k}\frac{(n-1)!}{(n-1-k)!}\frac{\partial^{n-k} f}{\partial z^{n-k}}B_{k|l}(f(z); z)\partial^{l+1}\phi(f(z)) \\
&=\sum_{k=1}^{n} B_{n|k}(f(z); z)\partial^k\phi
\end{aligned} \tag{7.10}$$

are defined by the conformal transformation for $\partial^n\phi$. Here $B_{n|k}$ are the incomplete Bell polynomials in the derivatives (expansion coefficients) of $f$ and $S_{n_1|n_2}(f; z)$ are the generalized higher-derivative Schwarzians. To calculate $S_{n_1|n_2}$ we cast the normal ordering according to:

$$\begin{aligned}
&\frac{1}{n_1!n_2!} : \partial^{n_1}\phi\partial^{n_2}\phi : (z) = lim_{\epsilon\to 0}\{\partial^{n_1}\phi(z+\frac{\epsilon}{2})\partial^{n_2}\phi(z-\frac{\epsilon}{2}) \\
&+\frac{(-1)^{n_1}(n_1+n_2-1)!}{n_1!n_2!\epsilon^{n_1+n_2}}\}
\end{aligned}$$

(again, the regularization parameter $\epsilon$ here is not to be confused with $i\epsilon$ for the location of $U_\alpha$ and/or infinitezimal transformation parameter $\epsilon(z)$). Under the conformal map $z \to f(z)$, this expression transforms according to

$$\begin{aligned}
&\frac{1}{n_1!n_2!} : \partial^{n_1}\phi\partial^{n_2}\phi : (z) \\
&\to \frac{1}{n_1!n_2!}lim_{\epsilon\to 0}\{\sum_{k_1=1}^{n_1}\sum_{k_2=1}^{n_2} B_{n_1|k_1}(f(z+\frac{\epsilon}{2}); z)B_{n_k|k_2}(f(z-\frac{\epsilon}{2}); z) \\
&\times(\partial^{k_1}\phi(z+\frac{\epsilon}{2})\partial^{k_2}\phi(z-\frac{\epsilon}{2}) + \frac{(-1)^{k_1}(k_1+k_2-1)!}{(f(z+\frac{\epsilon}{2})-f(z-\frac{\epsilon}{2}))^{k_1+k_2}} \\
&+\frac{(-1)^{n_1}(n_1+n_2-1)!}{n_1!n_2!\epsilon^{n_1+n_2}})\}
\end{aligned}$$

Expanding in $\epsilon$, we extract (upon cancellations of the divergent terms) the higher-order Schwarzians to be given by:

$$S_{n_1|n_2}(f;z) = \frac{1}{n_1!n_2!}\sum_{k_1=1}^{n_1}\sum_{k_2=1}^{n_2}\sum_{m_1\geq 0}\sum_{m_2\geq 0}$$

$$\sum_{p\geq 0}\sum_{q=1}^{p}(-1)^{k_1+m_2+q_2-m_1-m_2}(k_1+k_2-1)!$$

$$\times\frac{\partial^{m_1}B_{n_1|k_1}(f(z);z)\partial^{m_2}B_{n_2|k_2}(f(z);z)B_{p|q}(g_1,...,g_{p-q+1})}{m_1!m_2!p!(f'(z))^{k_1+k_2}}$$

$$g_s = 2^{-s-1}(1+(-1)^s)\frac{\frac{d^{s+1}f}{dz^{s+1}}}{(s+1)f'(z)}; s = 1,...,p-q+1 \tag{7.11}$$

with the sum over the non-negative numbers $m_1, m_2$ and $p$ taken over all the combinations satisfying

$$m_1 + m_2 + p = k_1 + k_2.$$

Here, in (7.10) and (7.11) $B_{n|k}(g_1, ...g_{n-k+1})$ are the incomplete Bell polynomials.

Note that, just as the ordinary Schwarzian satisfies the well-known composite relation for any combination of conformal transformations $z \to f(z) \to g(f)$:

$$S_{1|1}(g(z);z) = (\frac{df}{dz})^2 S_{1|1}(g(f);f) + S_{1|1}(f(z);z),$$

the generalized Schwarzians $S_{n_1|n_2}$ also satisfy the composite relations

$$S_{1|1}(g(z);z)$$
$$=\sum_{k_1=1}^{n_1}\sum_{k_2=1}^{n_2} B_{n_1|k_1}(f(z);z)B_{n_2|k_2}(f(z);z)S_{k_1|k_2}(g(f);f)$$
$$+S_{n_1|n_2}(f(z);z).$$

(see also [119, 120] who considered the alternative types of higher-order Schwarzians in a rather different context).

Now let us apply the same procedure to the irregular vertex operators $U_\alpha$ and $V_\beta$ in the partition-counting correlator (7.3). The straightforward computation of the infinitezimal transforms gives:

$$\delta_\epsilon V_\beta = [\oint \frac{du}{2i\pi}\epsilon(u)T(u); V_\beta(z)]$$
$$=: (\beta\partial\epsilon\partial\phi + \frac{1}{12}\beta^2\partial^3\epsilon)V_\beta : (z) \tag{7.12}$$

and

$$\delta_\epsilon U_\alpha(z) = [\oint \frac{du}{2i\pi}\epsilon(u)T(u); U_\alpha(z)]$$
$$= \sum_{n=1}^{\infty}\{: \frac{\alpha^n(\partial^n(\epsilon\partial\phi))}{n!(1-\frac{\alpha^n\partial^n\phi}{n!})}\prod_{N=1}^{\infty}\frac{1}{1-\frac{\alpha^N\partial^N\phi}{N!}}:$$
$$+ : \frac{\alpha^{2n}\partial^{2n+1}\epsilon}{(2n+1)!(1-\frac{\alpha^n\partial^n\phi}{n!})^2}\prod_{N=1}^{\infty}\frac{1}{1-\frac{\alpha^N\partial^N\phi}{N!}}:\}$$
$$+ \sum_{0\leq n_1<n_2<\infty} : \frac{\alpha^{n_1+n_2}\partial^{n_1+n_2+1}\epsilon}{(1-\frac{\alpha^{n_1}\partial^{n_1}\phi}{n_1!})(1-\frac{\alpha^{n_2}\partial^{n_2}\phi}{n_2!})}\prod_{N=1}^{\infty}\frac{1}{1-\frac{\alpha^N\partial^N\phi}{N!}}: \tag{7.13}$$

Integrating these infinitezimal transformations, we obtain the transformations of $U_\alpha$ and $V_\beta$ for the finite conformal transformation $f(z) = h(z)e^{-\frac{i}{z}}$. For $V_\beta$, we get

$$V_\beta(w)|_{w=i\epsilon} \rightarrow: e^{\beta\frac{\partial f}{\partial z}\phi(f(z))+\frac{\beta^2}{2}S_{1|1}(f(z);z)} : |_{f(z)=h(z)e^{-\frac{i}{z}}} \tag{7.14}$$

To determine the transformation law for $U_\alpha$, it is convenient to cast $U_\alpha$ as

$$U_\alpha = 1 + \sum_{N=1}^{\infty}\sum_{\{m_i\}}\alpha^N\prod_{n=1}^{N}(\frac{\partial^n\phi}{n!})^{m_n} \tag{7.15}$$

with the sum over $\{m_i\}; i=1,...,N$ being taken over all the combinations of non-negative $\{m_i\}$, satisfying

$$\sum_{n-1}^{N} nm_n = N$$

Now introduce the *exchange numbers* $\nu_{ij} \geq 0 (i,j=0,...,N)$ satisfying

$$\nu_{00} = 0$$
$$\nu_{ij} = \nu_{ji}$$
$$\nu_{jj} + \sum_{i=0}^{N}\nu_{ij} = m_j \tag{7.16}$$

in order to parametrize the internal normal ordering procedure for $U_\alpha$ as follows:

1. $\nu_{ij} = \nu_{ji}(i,j \neq 0)$ defines the number of internal couplings between $(\partial^i\phi)^{m_i}$ and $\partial^j\phi^{m_j}$ factors, creating internal singularities prior to the normal ordering;

2. $\nu_{ii}$ defines the number of intrinsic same-derivative couplings between $\partial^i\phi$'s inside each factor $(\partial^i\phi)^{m_i}$.

3. $\nu_{i0}$ counts the numbers of $\partial^i\phi$-operators left inside $(\partial^i\phi)^{m_i}$-block, that do not participate in the contractions.

Since each coupling between $\partial^i\phi$ and $\partial^j\phi$ contributes the factor $i!j!S_{i|j}(f;z)$ to the transformation law under $f(z)$, the overall transformation law for $U_\alpha$ is

$$\begin{aligned}U_\alpha(z) = \prod_n : \frac{1}{1-\frac{\alpha^n\partial^n\phi}{n!}} :\rightarrow \\ 1+\sum_{N=1}^{\infty}\sum_{q=1}^{N}\sum_{m_1,\ldots,m_q;\{\nu_{ij}\}}\alpha^N\prod_{p=1}^{q}\frac{(\delta_p(\phi;f))^{\nu_{p0}}m_p!(S_{p|p})^{\nu_{pp}}}{\nu_{p0}!(2\nu_{pp})!!} \\ \times \prod_{1\leq i<j\leq q}\frac{(S_{i|j})^{\nu_{ij}}}{\nu_{ij}!}\end{aligned} \tag{7.17}$$

where

$$\delta_p(\phi;f) = \sum_{k=1}^{p} B_{p|k}(f(z);z)\partial^k\phi(z) \tag{7.18}$$

Finally, since the Schwarzian of the conformal transformation $f(z)$ is nonzero, we need to account for the spontaneous breaking of the conformal symmetry by integrating the Ward identities, in order to match the partition-counting correlators $< U_\alpha V_\beta >$ in different coordinates. For that, we first have to integrate the infinitezimal "overlap" deformation of the correlator, emerging from the contraction of one of $\partial\phi$'s in the stress-energy tensor with $U_\alpha$ and another with $V_\beta$. The infinitezimal overlap deformation is given by the integral:

$$\begin{aligned}&\delta_{overlap} < U_\alpha(z)V_\beta(w) > |_{z=i\epsilon;w=0} \\ &= \sum_{N=1}^{\infty}\oint\frac{du}{2i\pi}\frac{\epsilon(u)}{(u-z)^{N+1}(u-w)^2} \\ &\times : \frac{\alpha^N\beta}{(1-\frac{\alpha^N\partial^N\phi}{N!})^2} \\ &\times \prod_{n=1;n\neq N}^{\infty}\frac{1}{1-\frac{\alpha^n\partial^n\phi}{n!}} : (z) : e^{\beta\partial\phi} : (w)|_{z=i\epsilon;w=0}\end{aligned}$$

$$= \sum_{N=1}^{\infty} (\partial_z^N [\frac{\epsilon(z)}{(z-w)^2}] + (-1)^{N+1}\partial_w [\frac{\epsilon(w)}{(z-w)^{N+1}}])$$

$$\times : \frac{\alpha^N \beta}{(1-\frac{\alpha^N \partial^N \phi}{N!})^2}$$

$$\times \prod_{n=1;n\neq N}^{\infty} \frac{1}{1-\frac{\alpha^n \partial^n \phi}{n!}} : (z) : e^{\beta\partial\phi} : (w)|_{z=i\epsilon; w=0} \tag{7.19}$$

This infinitezimal deformation is straightforward to integrate for the class of the conformal transformations (7.4). The overall integrated transformation for $U_\alpha$ under $f(z)$, with the overlap deformation included, is then given by

$$U_\alpha(z) = \prod_n : \frac{1}{1-\frac{\alpha^n \partial^n \phi}{n!}} :\to$$

$$(1 + \sum_{N=1}^{\infty} \sum_{q=1}^{N} \sum_{m_1,\ldots,m_q;\{\nu_{ij}\}} \alpha^N \prod_{p=1}^{q} \frac{(\delta_p(\phi;f))^{\nu_{p0}} m_p!(S_{p|p})^{\nu_{pp}}}{\nu_{p0}!(2\nu_{pp})!!}$$

$$\times \prod_{1\leq i<j\leq q} \frac{(S_{i|j})^{\nu_{ij}}}{\nu_{ij}!}) \prod_{n=1}^{\infty} \frac{1}{1-\frac{\alpha^n\beta}{n!}\frac{D_n(f(z);z)}{(1-\alpha^n\delta_n(\phi,f))}} \tag{7.20}$$

where

$$D_n(f(z);z) \equiv D(n) = -i\sum_{k=1}^{n}(-1)^{k+1}k!\frac{B_{n|k}(f(z);z)}{(f(z)-f(0))^k} \tag{7.21}$$

(to abbreviate notations, below we will also use the symbol $D(n)$ for $D_n(f(z);z)$). For the conformal transformations of the form $f(z) = h(z)e^{-\frac{i}{z}}$ the overall transformation law for $V_\beta(z)$ remains the same up to terms that vanish identically at $z = 0$:

$$V_\beta(w)|_{w=i0} \to e^{\beta\frac{df}{dz}\partial\phi(f(z))+\frac{\beta^2}{2}S_{1|1}(f(z);z)}|_{f(z)=h(z)e^{-\frac{i}{z}}} \tag{7.22}$$

Now that we are prepared to calculate the partition-counting correlator in the new coordinates, here comes the crucial part. The dipole's size in the new coordinates is

$$\beta\frac{df}{dz}|_{z\to 0} \to 0 \tag{7.23}$$

and shrinks to zero with our choice of $f(z)$. This drastically simplifies the calculation. While the operator $U_\alpha$ looks extremely cumbersome in the new coordinates (particularly, because of the complexities involving $\delta_n(\phi; f)$-operators), any contractions of derivatives of $\phi$ in $U_\alpha$ with $V_\beta$ bring down the factors proportional to $f'(z)$ and therefore vanish for the conformal transformations of the form (7.4). As a result, only the zero modes of $U_\alpha$ and $V_\beta$ contribute to the correlator in the new coordinates. Technically, this implies $\nu_{0j} = 0$ for all $j$. The correlator is then easily computed to give the generating function for the restricted partitions in terms of higher-derivative Schwarzians and incomplete Bell polynomials:

$$G(\alpha,\beta|\epsilon) \equiv < U_\alpha(z)V_\beta(w) > |_{z=i\epsilon,w=0} = e^{\frac{1}{2}\beta^2 S_{1|1}(f(w);w)}|_{w=0}$$
$$\times(1+\sum_{N=1}^{\infty}\sum_{q=1}^{N}\sum_{m_1,\dots,m_q;\{\nu_{ij}\}}\alpha^N\prod_{p=1}^{q}\frac{m_p!(S_{p|p})^{\nu_{pp}}}{(2\nu_{pp})!!}\prod_{1\leq i<j\leq q}\frac{(S_{i|j})^{\nu_{ij}}}{\nu_{ij}!})$$
$$\times\prod_{n=1}^{\infty}\frac{1}{1-\frac{\alpha^n\beta}{n!}D(n)}|_{z=i\epsilon} \tag{7.24}$$

The overall constant ($\epsilon$-independent) factor of $e^{\frac{1}{2}\beta^2 S_{1|1}(f;w)}|_{w=0}$ is related to the Casimir energy associated with the conformal transformation $f(w)$. It is irrelevant and disappears when the correlator is normalized with the inverse of the partition function of the system. The generation function for the partition numbers is then simply obtained by replacing this factor with 1.

Now the final step is to take the derivatives of $G$ in $\alpha,\beta$ and to $\epsilon$-order the result, retaining the finite terms as $\epsilon$ is set to 0. Straightforward calculation gives:

$$\lambda(N|Q) = (-i\epsilon)^{N+Q} : \sum_{N_1=0}^{N}\sum_{N_2=0}^{N-N_1}\sum_{\{m_j\geq 0\}}\sum_{\{n_j;p_j\geq 0\}}\sum_{\{\nu_{ij}\geq 0;\}}$$
$$\prod_{k=1}^{N_1}\prod_{q,r;1\leq q<r\leq N_1} : \frac{m_k!(S_{k|k})^{\nu_{kk}}(S_{q|r})^{\nu_{qr}}}{(2\nu_{kk})!\nu_{qr}!}$$
$$\times(D(n_1))^{p_1}\dots(D(n_Q))^{p_Q} :_{N+Q} \tag{7.25}$$

with the summations/products taken over the non-negative integer values of

$$N_1, N_2, N_3; m_1, \dots m_{N_1}; n_1, \dots n_Q; p_1, \dots p_Q, \{\nu_{ij}\}; 1\leq i,j\leq N_1$$

satisfying:

$$
\begin{aligned}
&N_1 + N_2 = N \\
&\sum_{j=1}^{N_1} j m_j = N_1 \\
&\sum_{j=1}^{s} n_j p_j = N_2 \\
&\nu_{ii} + \sum_{j=1}^{N_1} \alpha_{ij} = m_i; 1 \leq i \leq m_i \\
&\nu_{ij} = \nu_{ji}
\end{aligned} \tag{7.26}
$$

The $\epsilon$-ordering symbol : ... $:_{N+Q}$ in each monomial term of the sum $\sim: S...SD...D :_{N+Q}$ by definition only retains the terms of the order of $\epsilon^{-N-Q}$ upon the evaluation of each product $S...SD...D$, in order to ensure that the overall contribution is finite, upon multiplication by $(-i\epsilon)^{N+Q}$ (we refer to this procedure as $\epsilon$-ordering to distinguish it from the usual normal ordering defined for operators in CFT). Let us stress that, since each $S$ or $D$ has the finite and definite singularity order in $\epsilon$, the overall result for $\lambda(N|Q)$ is the exact analytic expression, given by the finite series, uniquely determined by the structures of $S$ and $D$ for each $f$ (with $f$ satisfying the constraints described above). This concludes our derivation of counting the restricted partitions, expressed in terms of finite series in the incomplete Bell polynomials and the generalized higher-derivative Schwarzians of the defining conformal transformation $f(z) = h(z)e^{-\frac{i}{z}}$.

## 7.3 Tests and Comments

Having presented the exact analytic expressionb for the number of the partitions, in this section we will provide some checks and examples of how the expression (7.25), constituting the main result of this section, works in practice. First of all, it is quite straightforward to demonstrate that the expression (7.25) leads to the correct answer for any partition number in the case of the conformal transformation $f(z)$ with the simplest choice $h(z) = 1$. Let us start from the most elementary case of the maximal length partition where obviously $\lambda(N|N) = 1$. for any $N$. Indeed, according to

(7.25), one has

$$\lambda(N|N) = (-i\epsilon)^{2N} : (D(1))^N :_{2N} = (-i\epsilon)^{2N}(-i)^N(-\frac{i}{\epsilon^2})^N = 1 \tag{7.27}$$

Similarly, it is easy to verify the case $1 = \lambda(N-1|N)$. Indeed,

$$\begin{aligned}\lambda(N|N-1) &= (-i\epsilon)^{2N-1} : (D(1))^{N-2}D(2) :_{N-1} \\ &=: (-i\epsilon)^{2N-1}(-\frac{i}{\epsilon})^{2N-4}(\frac{i}{\epsilon^3} - \frac{1}{2\epsilon^4}) + O(\epsilon) :_{2N-1} = 1\end{aligned} \tag{7.28}$$

Note (although irrelevant to our result) the appearance of the singular term $\sim \frac{1}{\epsilon}$ at this level (which was absent in the case of $\lambda(N|N)$). This term disappears upon the $\epsilon$-ordering procedure and is of no significance for our purposes, but the very reason for its emergence is also related to the Schwarzian singularities at $\epsilon \to 0$. It is easy to check that the results (7.27), (7.28) actually hold for any smooth regular $h(z)$ satisfying the conditions defined above, not just for $h = 1$. However, the case of $h = 1$ is the easiest one to verify the correctness of (3.22) for any partition, as this can be done by simple analysis of the $\epsilon$-dependence. Indeed, in general, each term for the partition number $\Lambda(N|Q)$ in (7.25) typically consists of $Q$ $D(n)$-factors and R $S$-factors (Schwarzians of all kinds and orders), where $R$ can in principle vary as $0 \leq R \leq [\frac{N-Q}{2}]$. However, in case of $h(z) = 1$ only the terms with $R = 0$ contribute. Indeed, the terms in each of the Schwarzians $S_{n_1|n_2}$, least singular in $\epsilon$ , are of the order of $\frac{1}{\epsilon^{n_1+n_2+2}}$ (e.g. $S_{1|1}(i\epsilon) = \frac{1}{12\epsilon^4}$. On the other hand, the $D(n)$-factors, consisting of combinations of incomplete Bell polynomials $B_{n|k}(f(z);z)$ with various $k$'s, have the lowest singularity order $\sim \frac{1}{\epsilon^{n+1}}$ for $h = 1$. Thus it is clear that each contribution with nonzero $R$ , upon multiplication by $(i\epsilon)^{N+Q}$ has the singularity order of at least $\frac{1}{\epsilon^R}$ and will disappear upon the normal ordering $: ... :_{N+Q}$. Furthermore, the only source of the lowest singularity terms in $D(n)$ of the order of $\frac{1}{\epsilon^{n+1}}$ is $B_{n|1}$, as all other $B_{n|k}$ with $k > 1$ are more singular, as it is easy to check. These terms stem from the derivatives $\partial^n(e^{-\frac{i}{z}})|_{z=i\epsilon}$ and are easily computed to be given by (skipping terms with the higher order singularities, not contributing to the $\epsilon$-ordering)

$$-\frac{i}{n!}(e^{\frac{i}{z}})\partial^n(e^{-\frac{i}{z}})|_{z=i\epsilon} = (\frac{i}{\epsilon})^{n+1} + h.s.$$

Thus each of the terms with $R = 0$:

$$\begin{aligned}&\sim (-i\epsilon)^{N+Q} : D(n_1)...D(n_Q) :_{N+Q} (N = n_1 + ... + n_Q) \\ &= (-i\epsilon)^{N+Q}(\frac{i^{N+Q}}{\epsilon^{N+Q}}) = 1\end{aligned}$$

contributes 1 to the sum. But the number of such terms obviously equals the number of partitions $\lambda(N|Q)$, hence this constitutes the proof that the formula (7.25) works correctly with the conformal transformation $f(z) = e^{-\frac{i}{z}}; h(z) = 1$. Although the case of $h(z) = 1$ is somewhat simplistic (e.g. with no Schwarzians entering the game), this by itself is already a non-trivial check of how the conformal invariance works in (7.25). Of course, with $h(z) \neq 1$ things change significantly and the Schwarzians of all orders contribute nontrivially to the expression (7.25) for the partitions. For example, consider $h(z) = cos(z)$ and $\lambda(4|2) = 2$. In this case, the Schwarzian $S_{1|1}$ is given by

$$6S_{1|1}|_{z=i\epsilon} = \frac{1}{2z^4} + \frac{2i}{z} = \frac{1}{2\epsilon^4} + \frac{2}{\epsilon} + O(\epsilon) \tag{7.29}$$

and does of course contribute to the normal ordering in general (the same is true for other $S_{n_1|n_2}$'s). According to (7.25) we have

$$\lambda(4|2) = (-i\epsilon)^6[:S_{1|1}(D(1))^2:_6 + :(D(2))^2:_6 + :D(1)D(3):_6] \tag{7.30}$$

Straightforward calculation gives:

$$\begin{aligned}
&(-i\epsilon)^6 : S_{1|1}(D(1))^2 :_6 = 0\\
&(-i\epsilon)^6 : (D(2))^2 :_6 = 1\\
&(-i\epsilon)^6 : D(3)D(1) :_6 = 1\\
&\lambda(4|2) = 2
\end{aligned} \tag{7.31}$$

For this partition, the Schwarzian related term still does not contribute, although its vanishing is not automatic but is related to the particular $\epsilon$-structure of the Schwarzian $S_{1|1}(f(z); z)$ for $h(z) = cos(z)$. For $\lambda(5|2)$ one calculates:

$$\begin{aligned}
&\lambda(5|2) = (-i\epsilon)^7[:S_{1|2}(D(1))^2:_7 + :S_{1|1}D(1)D(2):_7\\
&+ :D(1)D(4):_7 + :D(2)D(3):_7]\\
&(-i\epsilon)^7 : S_{1|1}D(1)D(2) :_7 = \frac{7}{12}\\
&(-i\epsilon)^7 : S_{1|2}(D(1))^2 :_7 = -\frac{1}{4}\\
&(-i\epsilon)^7 : D(1)D(4) :_7 = \frac{1}{4}\\
&(-i\epsilon)^7 : D(2)D(3) :_7 = \frac{17}{12}\\
&\lambda(5|2) = 2
\end{aligned} \tag{7.32}$$

so for this partition both $S$-type and $D$-type terms contribute nontrivially to $\lambda$. One can perform some similar tests to show that (7.25) works correctly. In general, however, the complexity of the manifest expressions for $\lambda(N|Q)$ grows dramatically with $N$ and especially with the difference $N-Q$, as not only the structure of higher order Schwarzians becomes increasingly cumbersome, but also the $\epsilon$-ordering procedure of the terms gets quite tedious. For this reason, the formula (7.25), although exact, is in practice less convenient for numerical computations of the partitions, compared to using the standard generating functions. Nevertheless, it casts the partition numbers in terms of exact finite analytic expressions in terms of the conformal transformation (7.4), which demonstrates the power of conformal symmetry and constitutes our main result.

# Bibliography


[1] M. Bianchi, V. Didenko, arXiv:hep-th/0502220
[2] A. Sagnotti, E. Sezgin, P. Sundell, arXiv:hep-th/0501156
[3] D. Sorokin, AIP Conf. Proc. 767 (2005) 172
[4] C. Fronsdal, Phys. Rev. D18 (1978) 3624
[5] S. Coleman, J. Mandula, Phys. Rev. 159 (1967) 1251
[6] R. Haag, J. Lopuszanski, M. Sohnius, Nucl. Phys B88 (1975) 257
[7] S. Weinberg, Phys. Rev. 133 (1964) B1049
[8] E. Fradkin, M. Vasiliev, Phys. Lett. B189 (1987) 89
[9] E. Skvortsov, M. Vasiliev, Nucl. Phys. B756 (2006) 117-147
[10] E. Skvortsov, J. Phys. A42 (2009) 385401
[11] M. Vasiliev, Phys. Lett. B243 (1990) 378
[12] M. Vasiliev, Int. J. Mod. Phys. D5 (1996) 763
[13] M. Vasiliev, Phys. Lett. B567 (2003) 139
[14] A. Bengtsson, I. Bengtsson, L. Brink, Nucl. Phys. B227 (1983) 31
[15] S. Deser, Z. Yang, Class. Quant. Grav 7 (1990) 1491
[16] A. Bengtsson, I. Bengtsson, N. Linden, Class. Quant. Grav. 4 (1987) 1333
[17] X. Bekaert, N. Boulanger, S. Cnockaert, J. Math. Phys. 46 (2005) 012303
[18] R. Metsaev, arXiv:0712.3526
[19] W. Siegel, B. Zwiebach, Nucl. Phys. B282 (1987) 125
[20] W. Siegel, Nucl. Phys. B 263 (1986) 93
[21] A. Neveu, H. Nicolai, P. West, Nucl. Phys. B264 (1986) 573
[22] T. Damour, S. Deser, Ann. Poincare Phys. Theor. 47 (1987) 277
[23] D. Francia, A. Sagnotti, Phys. Lett. B53 (2002) 303
[24] D. Francia, A. Sagnotti, Class. Quant. Grav. 20 (2003) S473
[25] D. Francia, J. Mourad, A. Sagnotti, Nucl. Phys. B773 (2007) 203
[26] J. Labastida, Nucl. Phys. B322 (1989)
[27] J. Labastida, Phys. Rev. Lett. 58 (1987) 632
[28] L. Brink, R.Metsaev, M. Vasiliev, Nucl. Phys. B586 (2000) 18
[29] X. Bekaert, S. Cnockaert, C. Iazeolla, M.A. Vasiliev, IHES-P-04-47, ULB-TH-04-26, ROM2F-04-29, FIAN-TD-17-04, Sep 2005 86pp.
[30] A. Campoleoni, D. Francia, J. Mourad, A. Sagnotti, Nucl. Phys. B815 (2009) 289-367

[31] A. Campoleoni, D. Francia, J. Mourad, A. Sagnotti, arXiv:0904.4447
[32] D. Francia, A. Sagnotti, J. Phys. Conf. Ser. 33 (2006) 57
[33] D. Polyakov, Int. J. Mod. Phys. A20 (2005) 4001-4020
[34] D. Polyakov, arXiv:0905.4858
[35] D. Polyakov, arXiv:0906.3663
[36] D. Polyakov, Phys. Rev. D65 (2002) 084041
[37] A. Fotopoulos, M. Tsulaia, Phys. Rev. D76 (2007) 025014
[38] I. Buchbinder, V. Krykhtin, Phys. Lett. B656 (2007) 253-264
[39] X. Bekaert, I. Buchbinder, A. Pashnev, M. Tsulaia, Class. Quant. Grav. 21 (2004) S1457-1464
[40] I. Buchbinder, A. Pashnev, M. Tsulaia, arXiv:hep-th/0109067
[41] I. Buchbinder, A. Pashnev, M. Tsulaia, Phys. Lett. B523 (2001) 338-346
[42] I. Buchbinder, E. Fradkin, S. Lyakhovich, V. Pershin, Phys. Lett. B304 (1993) 239-248
[43] I. Buchbinder, A. Fotopoulos, A. Petkou, Phys. Rev. D74 (2006) 105018
[44] G. Bonelli, Nucl. Phys. B 669 (2003) 159
[45] G. Bonelli, JHEP 0311 (2003) 028
[46] C. Aulakh, I. Koh, S. Ouvry, Phys. Lett. 173B (1986) 284
[47] S. Ouvry, J. Stern, Phys. Lett. 177B (1986) 335
[48] I. Koh, S. Ouvry, Phys. Lett. 179B (1986) 115
[49] A. Sen, JHEP 9912 (1999) 027
[50] A. Sen, B. Zwiebach, JHEP 0003 (2000) 002
[51] L. Rastelli, A. Sen, B. Zwiebach, JHEP 0111 (2001) 035
[52] M. Schnabl, Adv. Theor. Math. Phys. 10 (2006) no.4, 433-501
[53] M. Schnabl, JHEP 0702 (2007) 096
[54] T. Erier, M. Schnabl, JHEP 0910 (2009) 066
[55] L. Rastelli, A. Sen, B. Zwiebach, JHEP 0910 (2009) 066
[56] D. Polyakov, Phys. Rev. D65 (2002) 084041, arXiv:hep-th/0111227
[57] H Verlinde, Phys. Lett. B192 (1987) 95
[58] D. Polyakov, in progress
[59] E D'Hoker, D.H. Phong, Nucl. Phys. B636 (2002) 3-60, arXiv:hep-th/0110283
[60] E. D'Hoker, D.H. Phong, Nucl. Phys. B636 (2002) 61, arXiv:hep-th/0111016
[61] I.I. Kogan, D.Polyakov, Int. J. Mod. Phys. A18 (2003) 1827, arXiv:hep-th/0208036
[62] E. Fradkin, A.Tseytlin, Nucl. Phys B261 (1985) 1-27
[63] D. Friedan, E. Martinec, S. Shenker, Nucl. Phys. B271 (1986) 93
[64] S. Gubser, I. Klebanov, A.M. Polyakov, Phys. Lett. B428 (1998) 105-114
[65] J. Maldacena, Adv. Theor. Math. Phys. 2 (1998) 231-252, arXiv:hep-th/9711200
[66] J. Polchinski, L. Susskind, N. Toumbas, arXiv:hep-th/9903228
[67] I. Klebanov, A. Polyakov, Mod. Phys. Lett. A6 (1991) 3273-3281
[68] E. Witten, Nucl. Phys. B373 (1992) 187-213
[69] C. Pope, L. Romans, X. Shen, Nucl. Phys. B339 (1990) 191

[70] D. Knizhnik, Usp. Fiz. Nauk, 159 (1989) 401-453
[71] Z. Koba, H. Nielsen, Nucl. Phys. B12 (1969) 517-536
[72] E. Witten, Adv. Theor. Math. Phys. 2 (1998) 253
[73] J. Polchinski, Phys. Rev. Lett. 75 (1995) 4724
[74] N. Ohta, Phys. Rev. D33 (1986) 1681
[75] N. Ohta, Phys. Lett. B179 (1986) 347
[76] M. A. Vasiliev, Phys. Lett. B285 (1992) 225
[77] X. Bekaert, S. Cnockaert, C. Iazeolla, M. A. Vasiliev, arXiv:hep-th/0503128
[78] E.S. Fradkin, M.A. Vasiliev, Nucl. Phys. B291 (1987) 141
[79] E.S. Fradkin, M.A. Vasiliev, Phys. Lett. B189 (1987) 89
[80] M. A. Vasiliev, Nucl. Phys. B862 (2012) 341-408
[81] M. A. Vasiliev, Sov. J. Nucl. Phys. 32 (1980) 439, Yad. Fiz. 32 (1980) 855
[82] V. E. Lopatin and M. A. Vasiliev, Mod. Phys. Lett. A3 (1988) 257
[83] M. Vasiliev, Class. Quant. Grav. 30 (2013) 104006
[84] E.S. Fradkin and M.A. Vasiliev, Int. J. Mod. Phys. A3 (1988) 2983
[85] F. Berends, G. Burgers, H. Van Dam, Nucl. Phys. B260 (1985) 295
[86] D. Polyakov, Phys. Rev. D82 (2010) 066005
[87] D. Polyakov, Phys. Rev. D83 (2011) 046005
[88] D. Polyakov, J. Phys. A46 (2013) 214012
[89] A. Sen, B. Zwiebach, JHEP 0003 (2000) 002
[90] S. Giombi, I. Klebanov, arXiv:1308.2337
[91] T. Erler, JHEP 1311 (2013) 007
[92] T. Erler, JHEP 1104 (2011) 107
[93] T. Erler, M. Schnabl, JHEP 0910 (2009) 066
[94] M. Schnabl, Adv. Theor. Math. Phys. 10 (2006) 433-501
[95] T. Erler, M. Schnabl, JHEP 0910 (2009) 066
[96] M. Kroyter, Y. Okawa, M. Schnabl, S. Torii, B. Zwiebach, JHEP 1203 (2012) 030
[97] M. Kroyter, JHEP 1103 (2011) 081
[98] N. Berkovits, A. Sen, B. Zwiebach, Nucl. Phys. B587 (2000) 147-178
[99] N. Berkovits, Nucl. Phys. B450 (1995) 90
[100] N. Berkovits, JHEP 0004 (2000) 022
[101] I. Arefeva, P. Medvedev, A. Zubarev, Mod. Phys. Lett. A6, 949 (1991)
[102] I. Arefeva, A. Zubarev, Mod. Phys. Lett. A8 (1993) 1469-1476
[103] I. Arefeva, P. Medvedev, A. Zubarev, Nucl. Phys. B341 (1990) 464-498
[104] I. Bars, Y. Matsuo, Phys. Rev. D66 (2002) 066003
[105] I. Bars, I. Kishimoto, Y. Matsuo, Phys. Rev. D67 (2003) 066002
[106] I. Bars, arXiv:hep-th/0211238
[107] M. Douglas, H. Liu, G. Moore, B. Zwiebach, JHEP 0204 (2002) 022
[108] G.H. Hardy, S. Ramanujan, Asymptotic formulae in combinatory analysis, in Proceedings of the London Mathematical Society, 17 (1918) 75-115.
[109] H. Rademacher, *On the Partition Function p(n)*, in Proc. London Math. Soc. (2), 43:241-254, 1937
[110] P. Erdos, *On an elementary proof of some asymptotic formulas in the theory of partitions*, Ann. Math. (2). 43: 437-450.

[111] A. Belavin, A. Polyakov, A. Zamolodchikov, *Infinite Conformal Symmetry in Two-Dimensional Quantum Field Theory*, Nucl. Phys. B241 (1984) 333-380

[112] L. F. Alday, D. Gaiotto and Y. Tachikawa, *Liouville Correlation Functions from Four-dimensional Gauge Theories*, Lett. Math. Phys. **91** (2010) 167 [arXiv:0906.3219 [hep-th]].

[113] D. Gaiotto and J. Teschner, *Irregular singularities in Liouville theory*, JHEP **1212** (2012) 050 [arXiv:1203.1052].

[114] D. Gaiotto, *Asymptotically free N=2 theories and irregular conformal blocks*, J. Phys. Conf. Ser. **462** (2013) no.1, 012014 [arXiv:0908.0307].

[115] D. Polyakov, C. Rim, *Irregular vertex operators for irregular conformal blocks*, Phys. Rev. D93 (2016) no.10, 106002 [arXiv:1601.07756 [hep-th]].

[116] P. Di Francesco, P. Mathieu, D. Senechal, *Conformal Field Theory*, New York: Springer-Verlag (1997), DOI: 10.1007/978-1-4612-2256-9.

[117] E. Witten, *Noncommutative geometry and string field theory*, Nucl. Phys. B268 (1986) 253-294

[118] D. Polyakov, *Solutions in Bosonic string field theory and higher spin algebras in AdS*, Phys. Rev. D92 (2015) no.10, 106008 and the work in progress

[119] L. Bonora, M. Matone, *KdV equation on Riemann surfaces*, Nucl. Phys. B327 (1989) 415

[120] M. Matone, *Uniformization theory and 2-D gravity. 1. Liouville action and intersection numbers*, Int. J. Mod. Phys. A10 (1995) 289

[121] G. Almkvist, *A Hardy-Ramanujan formula for restricted partitions*, Journal of Number Theory 38 (1991) 135

[122] G. Almkvist, *Exact asymptotic formulas for the coefficients of nonmodular functions*, Journal of Number Theory 38 (1991) 145

[123] S. Govindarajan, N. Prabhakar, *A superasymptotic formula for the number of plane partitions*, arXiv:1311.7227

[124] E. Sezgin, P. Sundell, Nucl. Phys. B644 (2002) 303-370

[125] M. Vasiliev, Phys. Lett. B754 (2016) 187-194

[126] M. Vasiliev, Class. Quant. Grav. 30 (2013) 104006

[127] O. Gelfond, M. Vasiliev, Phys. Lett. B754 (2016) 187-194

[128] E. Fradkin, M. Vasiliev, Phys. Lett. B189 (1987) 89

[129] M. Vasiliev, Phys. Lett. B243 (1990) 378

[130] M. Vasiliev, Int. J. Mod. Phys. D5 (1996) 763

[131] X. Bekaert, N. Boulanger, S. Cnockaert, J. Math. Phys 46 (2005) 012303

[132] G. Bonelli, K. Maruyoshi, A. Tanzini, JHEP 1108 (2011) 056

[133] G. Bonelli, K. Maruyoshi, A. Tanzini, JHEP 1301 (2013) 014

[134] D. Gaiotto and J. Teschner, JHEP 1212 (2012) 050

[135] D. Gaiotto, J. Phys. Conf. Ser. 462 (2013) 012014

[136] E. Felinska, Z. Jaskolski, and M. Kosztolowicz, Math. Phys. 53 (2012) 033504

[137] D. Polyakov, C. Rim, Phys. Rev. D93 (2016) 106002

[138] H. Nagoya and J. Sun, J. Phys. A 43 (2015) 465203

[139] J. Gomis, B. Le Floch, JHEP 1603 (2016) 118

[140] T. Nishinaka and C. Rim, JHEP 1210 (2012) 138

[141] S. K. Choi and C. Rim, J. Phys. A49 (2016) 075201
[142] I. Arefeva, A. Koshelev, S. Vernov, Theor. Math. Phys. 148 (2006) 895-909, Teor. Mat. Fiz. 148 (2006) 23-41
[143] I. Arefeva, A. Koshelev, JHEP 0702 (2007) 041
[144] A. Koshelev, K. Sravan Kumar, P. Vargas Moniz, arXiv:1604.01440
[145] A. Sen, JHEP 0003 (2000) 002
[146] M. Schnabl, Adv. Theor. Math. Phys. 10 (2006) 433-501
[147] A. Sen, Phys. Scripta T117 (2005) 70-75
[148] L. Joukovskaya, JHEP 0902 (2009) 045
[149] M. Henneaux, S.-J. Rey, JHEP 1012 (2010) 007